Bacterial Starter Cultures for Foods

Editor

Stanley E. Gilliland

Professor

Animal Science Department

Oklahoma State University

Stillwater, Oklahoma

CRC Press

Taylor & Francis Group

Boca Raton London New York

CRC Press is an imprint of the

Taylor & Francis Group, an **informa** business

First published 1985 by CRC Press
Taylor & Francis Group
6000 Broken Sound Parkway NW, Suite 300
Boca Raton, FL 33487-2742

Reissued 2018 by CRC Press

Library of Congress Cataloging in Publication Data
Main entry under title:

Bacterial starter cultures for foods

Bibliography: p.
Includes index.
1. Bacterial starter cultures. 2. Food, Fermented.
3. Food additives. I. Gilliland, Stanley, E., 1940-
TP456.B32B33 1985 664'.024 84-15577
ISBN 0-8493-5686-5

A Library of Congress record exists under LC control number: 84015577

Publisher's Note
The publisher has gone to great lengths to ensure the quality of this reprint but points out that some imperfections in the original copies may be apparent.

ISBN 13: 978-1-315-89096-8 (hbk)
ISBN 13: 978-1-351-07006-5 (ebk)

Visit the Taylor & Francis Web site at http://www.taylorandfrancis.com and the
CRC Press Web site at http://www.crcpress.com

PREFACE

Since biblical times man has included cultured food products in his diet. Through the centuries the production of such cultured foods has been considered an art rather than a science. As a result of the different methods used by different groups of people throughout the world in producing these cultured foods there are many varieties available today. Even though man knew how to bring about the change in food resulting from the microbial action, for centuries it was unknown that this change was due to microorganisms. While man passed the knowledge of procedures for producing such food products down from generation to generation, today we still do not have complete knowledge concerning bacterial starter cultures from the standpoint of their growth and action in many food products. Much research has been devoted to studying bacterial starter cultures from the standpoint of their growth and action in foods, however, much yet appears to be unknown. The new era of biotechnology provides new opportunities for learning even more about bacterial starter cultures at the molecular level. Techniques involving such things as genetic engineering and plasma transfer offer the possibility of developing new and better bacterial starter cultures. The development of new food products resulting from the growth and action of bacterial starter cultures offers great opportunities in food research and development.

THE EDITOR

Stanley E. Gilliland is a professor in the Animal Science Department at Oklahoma State University, Stillwater, Oklahoma. He received the B.S. and M.S. degrees in Dairy Manufacturing from Oklahoma State University in 1962 and 1963, respectively. He was awarded the Ph.D. degree in Food Science from North Carolina State University, Raleigh, in 1966. After receiving the doctorate degree he served on the faculty in the Food Science Department at North Carolina State University with responsibilities for both research and teaching. His research activities involved the microbiology of milk products, focusing mainly on cultured milk products. In 1976, he joined the faculty in the Animal Science Department at Oklahoma State University. His responsibilities include both teaching and research. In 1978 he was recognized by the Oklahoma State University Chapter of Alpha Zeta as the outstanding teacher in the college of Agriculture. His research activities involve the microbiology of dairy products. Currently the major effort relates to the production of concentrated cultures of lactobacilli for use as dietary adjuncts and health or nutritional benefits that can be gained from their use. In addition to teaching and research activities, he is the departmental Graduate Programs Coordinator.

His research activities have resulted in 48 publications in various scientific journals and numerous papers presented at scientific meetings. Additionally he has been an invited speaker in a number of seminars, symposia, and lectures before both scientific and industry groups. In recognition of his research activities he was presented the Pfizer Award in Cheese and Culture Products Research by the American Dairy Science Association in 1979.

He is member of the American Dairy Science Association and has served as a member of the Editorial Board for the *Journal of Dairy Science*. He is a member of the Southern Division of the American Dairy Science Association for which he is serving as a director. He is also a member of the American Society for Microbiology in which he is the current Chairman of the Food Microbiology Division and has served as a member of the Editorial Board for *Applied Microbiology*. Other memberships include the Institute of Food Technologists, Council for Agricultural Science and Technology, Oklahoma Dairy Tech Society, Sigma XI, Gamma Sigma Delta, and Alpha Zeta. He currently serves as chairperson for a group of experts on the Indigenous Antibiotic and Bacteriostatic Systems in Milk and is a member of a group of experts on the Role of Cultured and Culture Containing Dairy Foods in Health for the U.S. National Committee of the International Dairy Federation.

CONTRIBUTORS

James N. Bacus
MicroSure, Inc.
Dallas, Texas

William L. Brown
President
ABC Research Corporation
Gainesville, Florida

Timothy M. Cogan
Principal Research Officer
The Agricultural Research Institute
Moorepark Research Centre
Fermoy, County Cork
Ireland

M. A. Daeschel
Microbiologist
USDA Food Fermentation Laboratory
Agricultural Research Service, Southern Region
Assistant Professor
Department of Food Science
North Carolina State University
Raleigh, North Carolina

H. P. Fleming
Research Leader
USDA Food Fermentation Laboratory
Agricultural Research Service, Southern Region
Professor
Department of Food Science
North Carolina State University
Raleigh, North Carolina

Stanley E. Gilliland
Professor
Animal Science Department
Oklahoma State University
Stillwater, Oklahoma

R. F. McFeeters
Research Chemist
USDA Food Fermentation Laboratory
Agricultural Research Service, Southern Region
Professor
Department of Food Science
North Carolina State University
Raleigh, North Carolina

Larry L. McKay
Professor
Department of Food Science and Nutrition
University of Minnesota
St. Paul, Minnesota

George W. Reinbold
President
Reinbold and Associates, Inc.
Wheat Ridge, Colorado

William E. Sandine
Professor
Department of Microbiology
Oregon State University
Corvallis, Oregon

T. Frank Sugihara
Consultant
Western Regional Research Center
U.S. Department of Agriculture
Berkeley, California

Randall K. Thunell
Director of Research
Northwest Culture Technology (CulTech)
Corvallis, Oregon

TABLE OF CONTENTS

Chapter 1

INTRODUCTION

Stanley E. Gilliland

TABLE OF CONTENTS

I. HISTORICAL DEVELOPMENT OF CULTURED FOODS

The discovery of the process of converting foods into new forms having different, yet desirable properties or characteristics, was probably accidental. In early biblical times certain foods, such as milk, were probably placed under conditions that resulted in an alteration of the products. The resulting altered food product likely possessed new characteristics including taste, aroma, texture, and appearance. Not only would such changes in the product have produced a new food product to add variety in the diet, but also a product that could be saved for later use. The new form provided a food that would not spoil as easily as the original raw product. Perhaps the most significant effect of such fermentations was the preservation of the food.

No knowledge existed at the time concerning bacteria, much less the specific ones involved in the food fermentations. The desired change was controlled, probably more or less by the manipulation of the raw products with regard to substances added, the container, and the temperature at which the product was held. Additionally, treatment of the product after the fermentation very likely added to the characteristics of the product. The method of preparing the product undoubtedly must have involved a certain amount of trial and error in selecting the method of handling that created the most desirable new product. The overall procedures probably varied from one group of people to another. This would have resulted in a wide variety of fermented foods. The converted foods resulting from the growth and action of microorganisms under rather crude conditions were the predecessors of the wide variety of cultured foods we have today.

II. DEVELOPMENT OF STARTER CULTURES

Through the years it was found that the fermentation process was improved by saving part of the fermented product to start the next batch. This represents the first use of starter cultures.

Cultured foods as we know them today, including such products as yogurt, buttermilk, hundreds of varieties of cheeses, many varieties of fermented sausages, and fermented vegetables and breads, probably all can be traced back to foods that were allowed to undergo a normal or natural fermentation under rather crude conditions. If we accept this, these fermentations would have occurred due to microorganisms included in the natural flora of the raw food product. The environment under which the raw product was placed (in addition to substances added to the raw food) would have resulted in conditions that selected for the desirable type of microorganism. Much variation probably occurred in the creation of fermented food products. The production of such cultured food products would be considered an "art" rather than a science.

The bacteria included in starter cultures of today in many cases include those bacteria that predominated in the historical fermented foods. Such cultures have been developed by isolating those same bacteria. Most cultured foods today are manufactured under sanitary and controlled conditions to help ensure that the desired bioconversion occurs in producing the converted food product. In many cases, especially with milk products, the food is exposed to treatment such as heat prior to adding the starter culture in order to reduce the number of undesirable microorganisms. This is done to ensure that the starter culture is able to produce the desired change in the product. In many cases, however, it is not possible to heat the product prior to fermentation so that competing bacteria may be destroyed.

The primary genera of bacteria included as starter cultures include *Streptococcus, Lactobacillus, Leuconostoc, Pediococcus,* and *Propionibacteria*. Most of these genera have one thing in common: they produce lactic acid during their growth in the food products (*Propionibacteria* are the exception). Thus for most cultured food products, lactic acid becomes a very important component. The result is that most cultured foods share one common characteristic — sourness or tartness with regard to taste.

Over the years, the starter culture industry has undergone tremendous development. More knowledge has developed concerning the starter culture bacteria involved in milk fermentations than for any of the other cultured foods. However, a great deal of information is currently being added concerning the microorganisms used in the manufacture of other cultured foods. Today, not only do we use specific strains or mixtures of strains of known species of bacteria as starter cultures, but we possess vast knowledge concerning factors influencing their activity. Activity is defined as the relative ability of the organism to rapidly carry out the desired change in the food product being cultured. The techniques for producing, handling, and using starter cultures has changed over the years. We are entering a new era where the role(s) of plasmids and genetic engineering are being investigated to learn more about the biological factors that control starter culture activity, and to develop new and improved strains that will more efficiently produce cultured food products. Many of these aspects will be discussed in subsequent chapters of this book.

Starter culture bacteria are very important in converting food into new products and exerting preservative actions on the food. Their influence with regard to exerting preservative actions involves antagonistic action toward other types of microorganisms. This aspect of starter culture activity has created interest in applying starter cultures to foods that traditionally do not undergo fermentation in order to enhance their preservation. In addition to their preservative action, considerable information has been collected indicating that starter cultures can provide certain nutritional and health benefits. These aspects are also discussed in subsequent chapters.

Chapter 2

THE STREPTOCOCCI: MILK PRODUCTS

William E. Sandine

TABLE OF CONTENTS

I. HISTORICAL

Streptococci, both the saprophytic and parasitic species, occur in milk because of their fastidious nature. Here, they find a source of carbohydrate (lactose), protein breakdown products, vitamins, lipids, and minerals, all of which they need for growth. The natural reservoir for members of the *Streptococcus* genus is green plant material and here these bacteria also find the nutrients needed for growth. Because of the close association of milk production with green plants, it is natural that streptococci would enter the milk supply, especially under the milk production conditions existent in earlier times. Hand milking of cattle into open containers easily resulted in the contamination of raw milk with a variety of organisms, including the streptococci. Along with members of the *Lactobacillus* genus, they played an important role in the natural souring of the milk important in preventing the growth of undesirable pathogens and spoilage bacteria; this ultimately led to our understanding of the use of certain selected lactic acid bacteria in the manufacture of the many fermented dairy products that are available today.

Because of their occurrence in milk, the streptococci were among the first genus of microorganisms to receive research attention by bacteriologists. The review articles by Sherman,[1] Hammer and Bailey,[2] Sandine et al.,[3] and Lundstedt[4] present the historical information summarizing many of the early studies done with these bacteria. Since this chapter concerns the streptococci in milk products, most of the information will be confined to the lactic streptococcal species. Members of the other three streptococcal groups (viridans, pyogenic, and enterococcus) also occur in raw milk and the enterococci are prevalent in certain fermented products, especially cheese.

II. TAXONOMY

From the studies of a number of investigators, particularly the early work of Storch,[5] Conn,[6] Weigmann,[7] Orla-Jensen,[8] Lister,[9] Hammer and Bailey,[2] Lancefield,[10] and Sherman,[11] we are able to place isolates of streptococci into one of the four groups mentioned above. Table 1 shows these groupings and lists the principal characteristics of the most important species present therein. All but the lactic group contain potentially pathogenic organisms and it is principally from this group that selected isolates are taken for use as starter cultures in the preparation of fermented dairy products. Table 2 shows the characteristics of the organisms in the lactic streptococcus group. Though they form a rather homogeneous group genetically,[12] the lactic streptococcal group contains three easily identifiable species or, as they are now called, subspecies.[13] These are *Streptococcus lactis* subsp. *lactis, S. lactis* subsp. *cremoris,* and *S. lactis* subsp. *diacetylactis*. These are usually referred to as *S. lactis, S. cremoris,* and *S. diacetylactis,* respectively.

Many other names for organisms that are legitimate members of the lactic group have been used by various scientists and appear in the literature. These have been given varietal designations such as "tardus" to refer to slow acid-producing types, "hollandicus" to refer to capsule-producing types, "maltigenes" to refer to malt-flavored strains,[20] and "aromaticus" to refer to strains with variable or weak abilities to ferment citric acid. It seems appropriate to refer to certain lactic streptococcal strains as varieties of the appropriate species when one wants to emphasize a particular stable characteristic of a designated strain.

Considerable confusion existed in the early days of research on the "streptococci" that fermented citric acid and produced diacetyl plus carbon dioxide. Two different genera were involved and the rather inert "associative organisms" often found in milk and fermented dairy products were at first proposed as belonging to the *Betacoccus*

Table 1
GROUPS OF STREPTOCOCCI

Group	Growth 10°C	Growth 45°C	Lancefield group	Example organisms
Pyogenic	−	−	A,C,H,F,I,B	*S. agalactiae* *S. pyogenes*
Viridans	−	+	?	*S. salivarius* *S. thermophilus* *S. bovis*
Enterococcus	+	+	D	*S. faecalis* *S. faecium* *S. durans*
Lactic	+	−	N	*S. lactis* *S. cremoris* *S. diacetylactis*

Table 2
DISTINGUISHING CHARACTERISTICS OF THE LACTIC STREPTOCOCCI

Characteristic observed	Results: *S. lactis*	*S. cremoris*	*S. diacetylactis*
Growth at pH 9.2	+	−	+ or −
Growth at 40°C	+	−	+ or −
Growth in 4% NaCl	+	−	+ or −
NH_3 from arginine[14]	+	−	+ or −
CO_2 formed in milk[15]	−	−	+
Diacetyl in milk[16]	−	−	+
Antibiotic[17]	Nisin	Diplococcin	+ (Unnamed)
Mol % G + C[3,12]	34—38	34—38	34—38
Type lactate produced[18]	L(+)	L(+)	L(+)
Serological group[10,19]	N	N	N

genus,[8] but were later placed in the *Leuconostoc* genus;[21] these bacteria are the subject of a separate chapter in this book. The lactic streptococcus called *S. diacetylactis*[15,22] is also a flavor-producing organism by virtue of its ability to ferment citric acid with the production of diacetyl, acetoin, 2,3-butylene glycol, and carbon dioxide.

III. FUNCTIONS

The lactose-fermenting and resultant acid-producing capability of lactic streptococci enable them to perform several important functions in milk and milk products. In cheese they concentrate and stabilize the curd by coagulating the protein and expelling moisture. They also prevent or discourage growth of undesirable spoilage and pathogenic bacteria by reducing the pH. They thereby contribute to the texture and can contribute either directly or indirectly to the formation of flavor compounds that are characteristic and expected in certain products. Thus, they perform essential functions in fermented foods. Because of this they have been exploited for thousands of years by man in contributing to food stability and flavor. It has only been relatively recently that we have come to a clear understanding of how these bacteria are able to perform these important functions.

IV. ISOLATION AND ENUMERATION

Lactic streptococci, especially *S. lactis* and *S. diacetylactis,* can be readily isolated from enrichment cultures of green plant material. Raw milk may also be used as a source, especially if the milk is obtained by hand milking into open containers. Green plant materials blended in milk and held at 25 to 30°C until coagulated will yield a large collection of streptococci of both the enterococcus and lactic groups. However, it is rare to be able to isolate *S. cremoris* by this procedure. This is probably because this bacterium is less competitive in nature than enterococci, *S. lactis,* or *S. diacetylactis* and therefore it is present in fewer numbers. Since *S. cremoris* occurs in lower numbers than the other two, diluting a sample on agar media to obtain isolated colonies results in its dilution to extinction and so it is difficult to isolate it from nature. In one study of 3000 isolates from 59 samples of commercial raw milk, only 4% were *S. cremoris.*[23] In another study by Hirsch,[24] 2 of 35 raw milk isolates were *S. cremoris,* and in another study of 31 producer samples,[25] only 3 yielded isolates of *S. cremoris.* Strains of *S. cremoris* are widely used in starter cultures for the manufacture of fermented dairy products, being the most desirable from a flavor standpoint,[26] and therefore they can be readily isolated from this source.

A number of studies have been conducted to develop plating media for lactic streptococci.[27] Some of these are general growth media and others are selective or differential media allowing the isolation of *S. diacetylactis,*[28] the differentiation between *S. lactis* and *S. cremoris,*[29,30] or the differentiation of acid-producing *S. lactis* and *S. cremoris* from aroma-producing *S. diacetylactis* and *Leuconostoc.*[31,32] A widely used general growth medium for lactic streptococci is M17 broth and agar.[33] The convenience of the methods now available to generate hydrogen and carbon dixide (BBL-Gas Pak® system, Baltimore Biological Laboratory, Cockeysville, Md.), has made it easy to create an anaerobic environment when incubating plates of lactic streptococci. This encourages better growth of the organisms and in at least one case has been shown to be essential for the proper differentiation between fast and slow strains of lactic streptococci.[34]

V. METABOLISM

A. Citrate

Many research articles concerned with the metabolism of citrate by lactic streptococci and *Leuconostoc* species have been discussed by Lawrence et al.[35] and Cogan.[36] While certain aspects of the metabolism between these two important flavor organisms are the same, important differences exist.[37-39] Figure 1 shows the pathway by which *S. diacetylactis* ferments citric acid resulting in the production of diacetyl and related compounds. Both *Leuconostoc* and *S. diacetylactis* have a specific permease to transport citrate across the cell membrane. This enzyme functions optimally at around pH 5[40] and since *Leuconostoc* are unable to ferment lactose rapidly due to their weak proteolytic ability, they must be grown in association with acid-producing lactic streptococci such as *S. lactis* and *S. cremoris* in order to ferment citric acid vigorously. *S. diacetylactis,* however, can ferment lactose at a rate sufficient to lower the pH such that they can utilize citric acid and produce diacetyl in single strain milk culture. Mutants of *S. diacetylactis* lacking the permease have been described which, though they contain citritase and other enzymes needed to synthesize diacetyl, produce only small amounts of diacetyl in milk.[41] Such strains may have been called *S. lactis* var. *aromaticus* in the past.

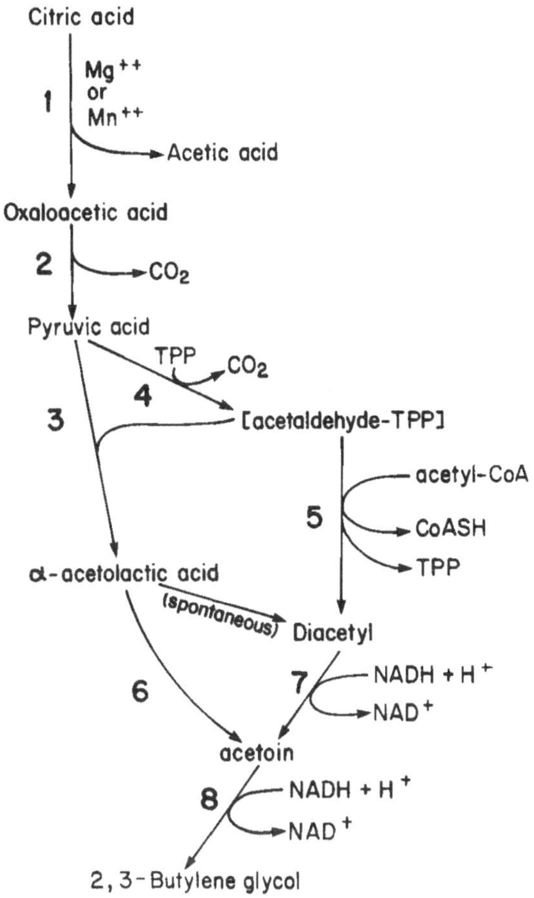

FIGURE 1. Citrate fermentation pathway occurring in *Streptococcus lactis* subsp. *diacetylactis*. Enzymes catalyzing the numbered reactions are 1, citrate lyase; 2, oxalacetate decarboxylase; 3, α-acetolactate synthase; 4, pyruvate decarboxylase; 5, diacetyl synthase; 6, α-acetolactate decarboxylase; 7, diacetyl reductase; 8, acetoin reductase.

B. Lactose

Lactose fermentation by lactic streptococci has been reviewed by McKay et al.[42] These organisms ferment lactose by converting it to glucose and galactose-6-phosphate by means of the enzyme phospho-β-galactosidase.[43] The glucose is then fermented through the hexose diphosphate pathway to produce lactic acid as the principal end product. The galactose-6-phosphate is utilized through the tagatose phosphate pathway for the production of additional lactic acid.[44] The pathways for the utilization of lactose and galactose are shown in Figure 2.

C. Glucose

On the basis of glucose fermentation, lactic acid bacteria are classified (after Orla-Jensen[8]) into homofermentative and heterofermentative types. Kandler[45] recently reviewed carbohydrate metabolism in these bacteria. The homofermentative types use the hexose diphosphate pathway and produce essentially all lactic acid from glucose. The heterofermentative types use the direct oxidative or 3/2 cleavage pathway wherein the glucose is decarboxylated and a pentose sugar is formed. Pentose is then cleaved

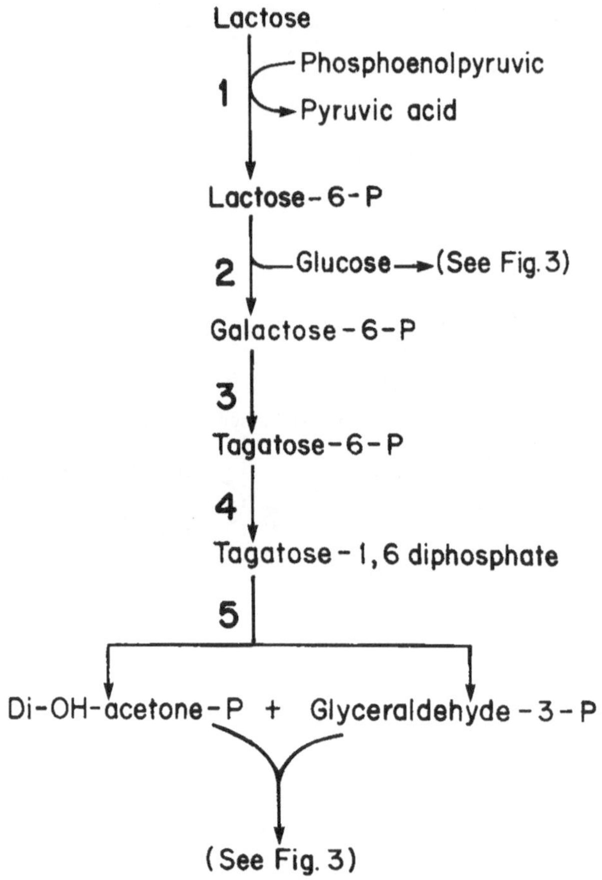

FIGURE 2. The tagatose pathway for galactose metabolism in lactic streptococci. Enzymes catalyzing the numbered reactions are 1, lactose specific enzyme (EII) of the PEP lactose phosphotransferase system; 2, phospho-β-galactosidase; 3, galactose-6-phosphate isomerase; 4, tagatose-6-phosphate kinase; 5, tagatose-1-6 diphosphate adolase.

into a C3 and C2 fragment, the C3 finally being reduced to lactic acid. The C2 portion may end up as acetate or be further reduced in the production of acetaldehyde and ethanol as may be required for regeneration of oxidized pyridine nucleotide. These reactions are summarized in Figure 3. Even homofermentative organisms have the enzymatic capability to be heterofermentative and thus the pathways and enzymes are under precise control, resulting in the formation of specific end products under certain conditions of growth. For example, when carbohydrate is limiting, homofermentative lactic streptococci can produce more heterofermentative end products.[46]

VI. GENETICS

From the discovery of DNA by the Swiss scientist Friedrich Miescher in 1869 and the discovery of its double helix structure in 1953 by Watson and Crick,[47] there followed a number of important basic research discoveries during the 1960s and 1970s. These discoveries led to the emergence of genetic engineering or recombinant DNA technology as we know it today. These basic discoveries included the finding that drug

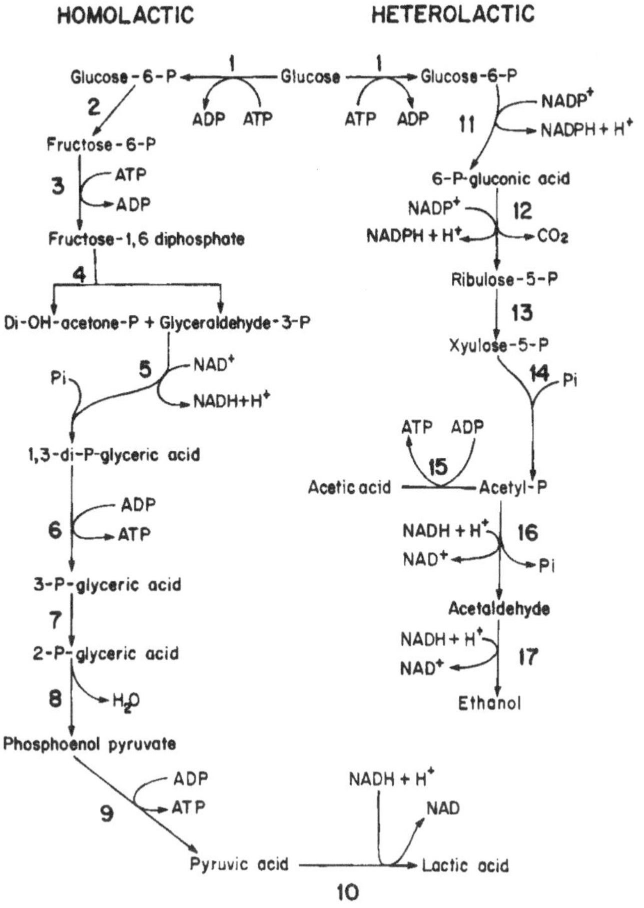

FIGURE 3. The homofermentative and heterofermentative lactic fermentations of glucose. The enzymes catalyzing the numbered reactions are 1, hexokinase; 2, phosphoglucose isomerase; 3, phosphofructokinase; 4, aldolase; 5, glyceraldehyde-3-phosphate dehydrogenase; 6, 3-phosphoglycerate kinase; 7, phosphoglyceromutase; 8, enolase; 9, pyruvate kinase; 10, lactic dehydrogenase; 11, glucose-6-phosphate dehydrogenase; 12, 6-phosphogluconate dehydrogenase; 13, ribulosephosphate-3-epimerase; 14, phosphoketolase; 15, acetokinase; 16, aldehyde dehydrogenase; 17, alcohol dehydrogenase.

resistance transfer factors were located on plasmids that are transferable from one bacterium to another. Another important finding was that restriction endonucleases that recognize the particular DNA sequence of four to six nucleotides that read the same forward and backward (palindromic sequences) will produce fragments with overlapping single-strand ends. These fragments can be aligned by complementary base pairing and joined to other DNA fragments generated by the same restriction enzyme. Also important was the isolation and purification of messenger RNA molecules so that DNA coding for their protein products could be synthesized using reverse transcriptase. Such DNA complementary to a specific messenger RNA is known as cDNA and can be used as a template to make the second strand (dsDNA) for use in cloning. Cloning is accomplished when cell and plasmid descendants from a single bacterium are constructed using these procedures.

These discoveries have ushered in a new era leading to advances in molecular biology with many potential applications in medicine and agriculture. The genetics of lactic acid bacteria promise to play an important role in these applications. Many new companies have been formed to conduct genetic engineering research with the hope of becoming profitable as a result of successful application of this technology in creating new products. Some useful products already made include a vaccine for hoof and mouth disease, insulin, a scours vaccine for pigs and calves, a porcine parvovirus vaccine, a cattle ward disease vaccine, and a bovine interferon vaccine against shipping fever. Other possibilities for cloning include the production of specific antibodies against cancer cells (some perhaps linked to cytotoxic chemicals), improved antibiotic production, the production of pure hormones in large quantities, other new vaccines, and the production of animal feed additives in large quantity.

Plasmids, which are units of DNA that replicate independent of the chromosome, are now known to exist in lactic acid bacteria. Their occurrence in both streptococci and lactobacilli is well established.[48-52] Isolation procedures are also well known now[53,54] though most of the plasmids at this writing are considered to be cryptic, i.e., the specific information coded in the plasmid DNA is as yet unknown. Agarose gel electrophoresis is used to screen strains of lactic acid bacteria for plasmid profiles. The profiles are quite characteristic for different strains and it has been suggested that the profiles can be used for strain identification.[48] While not simple, the development of such profiles could allow a bank of reference strains of known profiles to be established to serve as an international depository for strains of lactic acid bacteria. It has been suggested that there are only a few different strains of lactic streptococci from a bacteriophage sensitivity standpoint.[55,56] Perhaps the future publication of plasmid profiles for many strains will verify this contention.

Plasmids may occur in bacteria in a number of forms: monomeric, multimeric, open circular, closed circular, and linear. As a result agarose gel electrophoresis profiles may overestimate the number of plasmid species actually present in a particular strain. Improved techniques for plasmid isolation may minimize this problem.[54]

Lactobacilli have fewer plasmids than streptococci. Most strains so far studied have only one or two plasmids and many have none. In contrast the lactic streptococci may have up to 14 plasmids varying in molecular weight from 1 to 90 mdal. While most seem cryptic at this time it is known that plasmids code for certain dispensable functions in lactic streptococci.[50] These include lactose fermentation,[57,58] protease production,[51,58] citrate utilization,[59] resistance to inorganic salts,[60] and possibly nisin production.[61] The loss of plasmids carrying one or more of these functions leads to the production of descendant cells which are deficient in the corresponding ability or abilities. Little is known at present about factors that contribute to plasmid stability, but it is known that abusing lactic streptococci by over-incubation, freezing and thawing, and stressing with bacteriophages and antibiotics will increase the number of descendants in a culture that have lost certain plasmids.[62]

There are at least four different mechanisms by which genetic material may be transferred from one bacterium to another. These are transduction, conjugation, protoplast fusion, and transformation. Transduction is the transfer of genetic material mediated by a bacteriophage. While the process was first discovered in lactic streptococci as mediated by a virulent phage,[63] temperate phages that are unable to induce lysis of a host cell probably are the most important means by which surviving cells can be altered by this process. Figure 4 shows the possible outcomes when virulent and temperate viruses infect lactic streptococcal cells.[64] These processes not only can be demonstrated in the laboratory, but undoubtedly occur during milk fermentation with these bacteria.

Conjugation also has been demonstrated to occur in lactic streptococci.[65,66] This is a

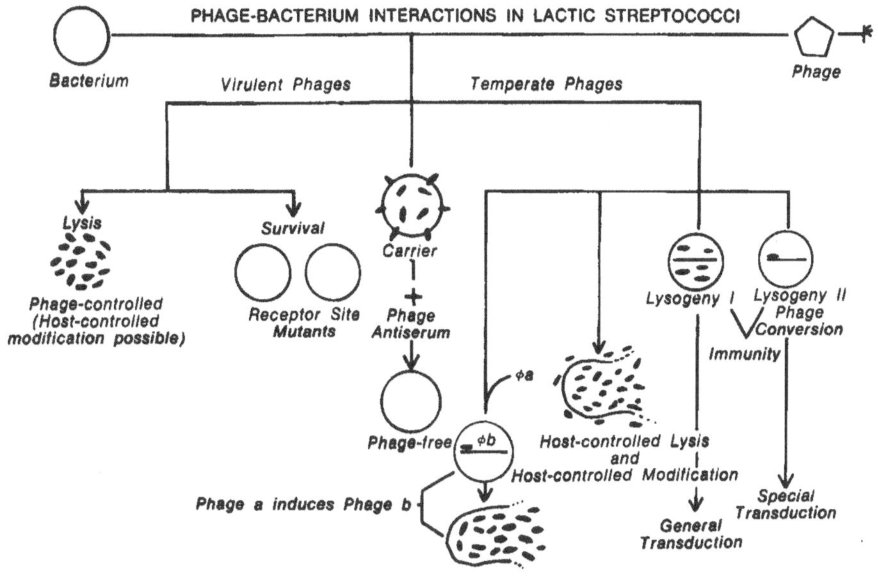

FIGURE 4. Possible bacteriophage-bacterium interactions in lactic streptococci.

process whereby appropriate cell types come together and genetic material moves in an oriented manner from a donor to a recipient through a conjugation bridge. This type of mating usually is facilitated on a solid surface such as an agar medium or a membrane filter.

Another method for genetic exchange is protoplast fusion.[67] This occurs when polyethyleneglycol (PEG) induces the fusion or joining together of protoplasts so that DNA from two cells are mixed and then there is opportunity for recombination events to occur. Following this, the protoplasts are regenerated on hypertonic medium and normal cells of a potentially new genetic make-up emerge. This type of genetic exchange would be expected to occur in nature only under rare and special conditions. However, use of this process does have the potential for strain improvement since it is likely that genetic information from diverse types of bacterial cells may interact and produce progeny cells with important new characteristics.

Transformation is the most recently discovered type of genetic exchange to occur among lactic streptococci.[68] In this type of exchange the DNA is isolated from a potential donor bacterium and mixed with a recipient bacterium. Under the appropriate conditions the recipient will take up the naked DNA, recombination events will occur, and progeny cells with characteristics provided by the donor DNA will emerge. Transformation has great potential for strain improvement, and perhaps transfer of such characteristics as immunity to bacteriophages, resistance to antibiotics, resistance to plasmid loss (insertion of plasmid markers within the chromosome), and less nutritive dependency may be possible.

Certain *Lactobacillus* plasmid genes have been cloned in *Escherichia coli*[69] and the transfer of the conjugative plasmid pAMβ from *S. lactis* to *Lactobacillus* strains also has been demonstrated.[70] This plasmid codes for resistance to erythromycin (macrolides), lincomycin, and spectrogramin (MLS phenotype). Other mechanisms of genetic transfer in lactobacilli have so far not been demonstrated. Undoubtedly, the resistance of these bacteria to lysis has resulted in less progress being made with them.

VII. FACTORS AFFECTING OPTIMUM PERFORMANCE

It is well known that a number of factors may affect (usually detrimentally) the performance of lactic streptococci during milk fermentation. A number of these will be discussed below.

A. Inhibitors in Milk

Inadequate growth by starter bacteria in milk may be due to the presence of antibiotics in the milk supply, residual sanitizers used in cleaning equipment, the presence of the antibiotic nisin produced by certain wild strains of *S. lactis* in the raw milk,[71] or an inadequate supply of protein breakdown products to support good growth.[72] Antibiotics (one or more of the penicillins most frequently used) may occur in the milk as a result of their use in the treatment of mastitis or udder diseases. Acceptable testing methods for antibiotics in milk are found in the most recent edition of the *Standard Methods for the Examination of Dairy Products;* immunological methods[73] as well as isotopic tracer dilution procedures (Charm test) have been described.[74] Cogan[75] has reported on the susceptibility of cheese starters to various antibiotics found in milk; additionally Cogan and Fitzgerald[76] described a test for detecting penicillin milk. Research on detection methods for antibiotics in raw milk will no doubt continue since there is great demand for a quick, accurate test that can be used at the farm level.

B. Strain Compatibility

Maintenance of mixed strain starters in cheese factories is not practiced much any longer, partially because repeated subculture of mixtures of strains of lactic streptococci may result in the decrease in numbers or loss of all but one of the strains so that essentially a single strain culture results. Factors contributing to the dominance or overgrowth of a mixed culture by one strain include differences in generation times, acid sensitivities, the production of antibiotics by the component strains,[77] differences in optimum temperature, and rates of plasmid loss.

The rationale behind using mixtures of strains in starter cultures is that in the event one strain becomes infected by a specific bacteriophage, the other strains will be unaffected and enable the culture to continue to produce acid during bulk culture preparation or cheesemaking. In four studies, milk cultures initially inoculated with equal volumes or numbers of single strains were dominated by one of the strains after only a few subcultures.[78-81] In each study imbalances were detected by phage infection of the multiple strain starter. The strain whose elimination by phage infection caused the greatest inhibition of acid production was said to dominant. Recently, another approach in dominance studies has been used: the employment of streptomycin-resistant mutants.[82] This has also revealed interactive effects between individual strains in mixed strain starters. Another dominance contributor is different rates of survival after thawing when frozen concentrates are used either to directly inoculate the cheese vat milk or to inoculate the bulk starter medium. In this regard, Leach and Sandine[83] have shown that differential survival and recovery from freezing are factors in production and alteration of dominance patterns in mixed strain starter cultures.

Recent trends toward the use of fewer but defined strains without rotation in bulk starter, cheese, and cultured buttermilk manufacture will undoubtedly result in more studies on strain dominance since mixtures of specific strains will be used in most processing plants.[84-86] This type of starter culture system will justify careful studies to learn more about the interactive effects of individual strains when they are growing with other individual strains.

C. Temperature

Both *S. lactis* and *S. cremoris* will vary for optimum growth temperature depending upon individual strains of each species. This can be a factor in strain dominance. From the work of Collins[87] it is clear that the best temperature for use in incubating a mixture of strains in the preparation of a bulk starter culture is 27°C. Use of this temperature will minimize dominance effects on individual component strains which may have differing optimum temperatures. While the optimum temperature for lactic streptococci is generally regarded to be 30°C, this is too high for a number of strains.

D. pH

While milk is one of nature's most nearly perfect foods for man and bacteria, it needs improvement for use as a bulk starter medium. Lactic streptococci produce more than 10% of their weight per minute in lactic acid. Strangely enough this fact has largely been ignored, particularly on the commercial scene for most of the years that bulk starter has been prepared to initiate milk fermentations. The growth of lactic starters to prepare frozen concentrates has been conducted under continuous neutralization for a number of years, but the application of this principle whether it be external or internal pH control has only been done recently in the industry.[88] Richardson et al.[89] pioneered the work on externally pH-controlled whey media for preparation of bulk cultures while Willrett et al.[90] and Sandine and Ayres[64,91] introduced the concept of preparation of bulk cultures using media with internal pH control. Other pH control methods have been used such as that of Osborne and Mabb[92] who developed a diffusion culture technique which allows for the removal of toxic end products of metabolism from starter cultures by dialysis. Also, Swedish workers have overcome the problem associated with the high lactose concentration of milk by developing a method to prepare starter culture concentrates in low lactose milk.[93] This eliminates the possibility that cultures in the maximum stationary phase will continue to produce acid from lactose to their own detriment. Lowering the lactose content eliminates this problem. The lactose is removed from the milk by Sephadex® column percolation and skim milk, then is added back to provide a final concentration of about 2% lactose.

The work of Harvey[94] is usually cited as the reason for controlling the pH drop in bulk starter media. When strains were allowed to grow so that the pH fell below 5, considerable injury to the cells resulted. Therefore, in the above techniques and media cited, care is taken to prevent this detrimental effect of low pH. With externally pH-controlled whey medium, the pH is maintained at 6.2;[89] for internally pH-controlled medium such as PHASE 4, the pH drop is controlled by an insoluble buffer so that the lactic acid produced is immediately neutralized, though the pH does fall to around 5.15 to 5.2.[90] Studies with both these types of media also have shown that in the unlikely event the bulk starter medium becomes contaminated with pathogens such as *Staphylococcus aureus* early in the preparation of the bulk culture the starter streptococci will out-compete the staphylococci and prevent their growth.[89] A starter medium for Italian-type cheese also has been described that uses the internal pH-control principle.[95]

E. Growth Media

With the exception of *S. bovis,* all of the streptococci that occur in nature appear to be quite fastidious in their growth requirements. Most will grow well in milk and since pathogenic streptococci such as *S. pyogenes* are associated with a milk-producing environment, justification for pasteurization is again seen.

The media used in the laboratory to cultivate the lactic streptococci are quite complex. The most widely used are MRS medium,[96] M17 medium,[33] and lactic or Elliker medium.[97] The latter recently was heralded in *Current Contents*[98] since the article describing this medium has been cited a record number of times according to Science Citation Index® and therefore is a Citation Classic®. M17 medium also is an excellent growth medium and is widely used for cultivating lactic streptococci. It also is suitable for use in the plaquing assay for bacteriophages. It contains β-disodium glycerophosphate as a buffer that neutralizes much of the acid produced by the growing organisms and allows for higher cell populations and larger colonies to be achieved on agar plates.

Several plating media have been described for the differentation among the three subspecies of lactic streptococci, i.e., *S. lactis, S. cremoris,* and *S. diacetylactis*. The reader is referred to a recent issue of the *Compendium of Methods for the Microbiological Examination of Foods*[27] for the procedures and composition of these different media. Another type of medium in which there has been considerable interest for the lactic streptococci has been one to differentiate between fast- and slow-growing cultures. Slow cultures are those that fail to coagulate nonfat milk when incubated at 21 to 22°C for 18 hr. Cultures that are able to coagulate milk when inoculated at the 1% level and so incubated are termed fast.[36] The work of Citti et al.[99] and Pearce et al.[100] made it clear that at least in some cases, fast cultures are more proteolytic than slow and therefore are able to obtain the needed nitrogen from milk protein and reach higher populations, thereby coagulating milk. Slow cultures, because of their limited proteolytic ability, are unable to reach sufficient populations to do this. Limsowtin and Terzaghi[101] have described a milk-based medium for distinguishing between fast and slow culture but it is of limited use because it does not work for all strains. Huggins and Sandine,[34,102] and Willrett[103] have described media called FSDA I and FSDA II, respectively, that distinguish between fast and slow strains. The composition for these media and procedures for use are shown in Tables 3 and 4.

In considering the growth of lactic streptococci in milk, it must be pointed out that at certain seasons of the year more inoculum may be needed for cheesemaking in order to achieve the same type of acid production as is experienced with lower inoculation rates at other times of the year. Late lactation milk and milk produced as the winter season approaches is usually deficient in certain factors that allow best growth of the lactic streptococci.[72] While it appears obvious to add certain stimulants to milk that might overcome this deficiency, economic and state and federal regulatory considerations have so far not allowed this.

F. Bacteriophages

Most authors in introducing papers dealing with research on bacteriophages point out that these agents are the single most important cause of slow or inhibited acid production by lactic starter streptococci in the commercial environment.[104] Because of this, many special precautions and specially formulated starter culture media have been developed over the years in an effort to control this serious problem. A number of publications have reported these developments,[35,105-108] but the most promising solution to this problem appears to be use of phage-resistant mutants.[106,109-111] The use of defined single strains and their phage-resistant mutants is now enjoying considerable success in the U.S. in Cheddar and Monterey Jack cheese operations and in cottage cheese and cultured buttermilk manufacture.[86,112] In Europe certain Cheddar cheese plants, especially in Ireland, also are using this approach.[113] The economic impact of this culture system also has been described and it would appear that within a few years most of the progressive cheese plants in the world that are making these types of cheese will be using some kind of a defined strain program without any culture rotations.[84]

More information on lactic bacteriophages can be obtained from the recent article by Teuber and Lembke.[114]

Table 3
FAST-SLOW DIFFERENTIAL AGAR
(FSDA I)

Ingredient	Amount
Agar	10 g
β-Disodium glycerophosphate	19 g
Litmus	1 g
Nonfat milk powder	100 g
Distilled water	1 ℓ

Note: This medium is prepared as two separate components, A and B. Component A is made by dissolving 10 g of Davis agar (Davis Gelatine Ltd., Christchurch, New Zealand) and steaming 30 min in a 2-ℓ flask containing 550 mℓ of distilled water, followed by addition and mixing of 19 g of sodium glycerophosphate (Ruger Chemical Co., Princeton, N.J.) and 1.0 g of Bacto litmus (DIFCO, Detroit, Mich.). Component B is made by dissolving 100 g of inhibitor-free nonfat dry milk powder in 450 mℓ of distilled water in a 1-ℓ flask. Components A and B are sterilized separately by autoclaving for 17 min at 121°C. The sterilized components are quickly cooled to 55°C in a water bath, mixed together by gently swirling in a larger flask, poured (20 mℓ) into Petri plates, and briefly flamed with a Bunsen burner to eliminate bubbles. Plates are allowed to dry in an inverted position for at least 24 hr at 30°C or for 48 hr at 20 to 25°C. The predried plates are inoculated by streaking or spreading 0.1 mℓ aliquots of appropriately diluted cultures on the surface of the agar. The inoculated plates then are incubated anaerobically (Gas Pak®, Baltimore Biological Laboratory, Cockeysville, Md.) at 30°C for 48 hr. Fast (Lac^+, Prt^+) colonies are 1 to 3 mm in diameter, shiny, white, convex, and surrounded by a red zone against the blue background of the medium indicating lactose fermentation. Slow colonies are 0.2 to 0.5 mm in diameter, translucent, and flat.

G. Capsule Production

On occasion one may isolate from raw milk or dairy products lactic streptococci which when grown in milk will produce a ropy, thick coagulum. The ropy character is due to capsule or polysaccharide production. Investigators have taken advantage of this for many years to produce the thick-bodied milk products that are characteristic of the Scandinavian countries. These are called vilia, filia, and viila. Organisms that produce this capsular material are called *S. lactis* var. *hollandicus* or *S. cremoris* var. *hollandicus*. Recently the induction of slime formation in lactic streptococci by the use of *p*-fluorophenylalanine has been described.[115]

H. Storage

For the routine maintenance of acid-producing streptococci in the laboratory, it is important not to store the mature cultures in the presence of the acid they have pro-

Table 4
FAST-SLOW DIFFERENTIAL AGAR (FSDA II)

Ingredient	Amount
Agar	10 g
Bromcresol purple (0.25% w/v solution)	20 m*l*
Nonfat milk powder	100 g
Trimagnesium phosphate	5 g
Distilled water	1 *l*

Note: This medium is prepared in three separate components A, B, and C. Component A is made by adding 20.0 m*l* of a 0.25 (w/v) solution of bromcresol purple and 10.0 g of Davis agar (Davis Gelatine, Ltd., Christchurch, New Zealand) to 400 m*l* of distilled water. Component B is prepared by dissolving 100 g of inhibitor-free nonfat milk powder in 500 m*l* of distilled water. Component C is made by adding 5.0 g of trimagnesium phosphate (Stauffer Chemical Co., St. Louis, Mo.) to 100 m*l* of distilled water; this component can be made in multiple units and stored at room temperature. Components A and C are autoclaved at 121°C for 10 min. The three components are then tempered to 55°C in a water bath and components B and C are added aseptically to component A. The complete medium is then mixed on a magnetic stirrer and further cooled to 45°C in a water bath. Plates are poured and allowed to air dry for 48 hr at 22 to 25°C. Occasional swirling of the flask is necessary to maintain the insoluble magnesium phosphate buffer in suspension while pouring plates. Appropriately diluted (0.1 m*l* aliquots) samples are smeared on the surface of the predried plates which are incubated anaerobically (Gas Pak®, Baltimore Biological Laboratory, Cockeysville, Md.) at 30°C for 48 hrs. The fast Lac^+, Prt^+ wild type colonies are fully colored yellow colonies against an opaque baby-blue background. The Lac^+, Prt^- colonies are less brightly colored and resemble yellow doughnuts. Colonies of Lac^-, Prt^+, and Lac^-, Prt^- are indistinguishable from each other, but are clearly differentiated from the other two types; they appear as colorless, translucent colonies.

duced during growth. Storage under these conditions will result in cellular injury and promote the loss of plasmids which may increase the proportion of slow cells in the population.[62] Such cultures will become slow and no longer proper for use in milk fermentations. Therefore, it is important in the maintenance of such cultures in milk that they be transferred to fresh milk and stored refrigerated without incubation. When a new culture is needed it is removed from the refrigerator, incubated, transferred to fresh medium, and placed in the refrigerator again. Mature cultures, however, may be stored at 2 to 5°C in stabs of M17 agar medium.[33] Cultures also can be frozen and stored at −40°C or below in the freeze-dried state. Both freezing and lyophilization are detrimental to cells and may, depending upon the strain, cause destruction of as much as 90 to 95% of the cells. This high rate of cell death must be overcome through the use of cryoprotective agents such as glycerol, if the frozen cultures are going to be used in directly inoculating cheese vat milk or bulk starter media. In the latter case growing and freezing the bacteria in internally pH-controlled media has been shown to be beneficial.[116] Frozen storage in liquid nitrogen (−196°C) also affords stability to cultures, but storage in mechanical freezers at ultralow dry air temperatures such as −40°C and below is nearly as good and the difficulties of working with liquid nitrogen are then avoided.

I. Variation

Changes occur in certain important characteristics of the lactic streptococci from time to time, therefore, they should be maintained under controlled conditions to minimize these changes.[117] Stress conditions that contribute to variations in starter cultures include the presence of antibiotics, incubation below or above their optimum growth temperatures, freezing and thawing, and growth in a nutritionally inadequate medium. Changes that may occur under stress are the appearance of a ropy characteristic, loss of fast acid production, and the production of abnormal flavor compounds such as certain aldehydes that have a malty aroma. Limsowtin et al.[118] have described the heterogeneity that exists in a lactic starter streptococci from the standpoint of acid production and bacteriophage sensitivity. It is likely that this heterogeneity may be manifested for many other characteristics, but adequate testing procedures to measure these differences are not yet available.

While the genetic material of lactic streptococci may not be nearly as diverse as the genetic material in complex, highly differentiated organisms, it is nonetheless unreasonable to expect that every cell in a population is like every other cell. One may think of cultures as being homogeneous because repeatable results are obtained in biochemical tests; however, there likely are many heretofore unmeasured differences between individual cells in a microbial population and occasionally manifestations of these are seen in culture variability.

J. Testing Strains

Because of the diverse capabilities of different lactic streptococci and the inherent variation within them, it is important to be able to test them for different characteristics. Sandine[108] has recently described most of the testing procedures that are commonly used for these bacteria. These include different tests for acid-producing ability, flavor or diacetyl production, gas production, bacteriophages, lysogeny, and for agglutination. Another simpler test for agglutination has been described more recently.[119]

VIII. CONCLUSIONS

Because of their importance in the production of many fermented foods, the lactic streptococci will continue to be studied extensively by many research investigators. This was emphasized not long ago at the meetings in The Netherlands on "Lactic Acid Bacteria in Foods"; over 200 scientists attended from most of the European and Western countries and many were just beginning new research on lactic acid bacteria. So while we have gained considerable knowledge about these organisms, scientists realize that there is much more to be learned. Their role in human and animal health still remains to be more fully understood. Also, successful recombinant DNA technology with these organisms, not only to improve their usefulness in fermentation, but to enable their use as vehicles for generation of products useful to man will no doubt be important developments of the future.

REFERENCES

1. Sherman, J. M., The streptococci, *Bacteriol. Rev.*, 1, 3, 1937.
2. Hammer, B. W. and Bailey, D. E., The volatile acid production of starters of organisms isolated from them, *Iowa Agric. Exp. Stn. Bull.*, 55, 223, 1919.
3. Sandine, W. E., Radich, P. R., and Elliker, P. R., Ecology of the lactic streptococci, a review, *J. Milk Food Technol.*, 35, 176, 1972.

4. Lundstedt, E., Some reflections on the development of starters for cultured dairy products industry, *Cult. Dairy Prod. J.*, 18, 10, 1983.
5. Storch, W., Untersuchungen über Butterfehler and Soüerung des Rahms, *Milch Ztg.*, 19, 304, 1890.
6. Conn, H. W., Bacteria in Milk, Cream and Butter, 2nd Annu. Rep., Storrs School, Storrs, Conn. 1889, p. 13.
7. Weigmann, H., Über den jetzigen stand der bakteriologis chen Forshung aug dem Gebeite der Kasereifingprozesses, *Cent. Bakteriol.*, 2(2), 150, 1896.
8. Orla-Jensen, S., *The Lactic Acid Bacteria*, (2nd ed. 1942), Copenhagen, I Komission Hos Ejnar Munksgaard, 1919, 196.
9. Lister, J., On the lactic fermentation and its bearing on pathology, *Trans. Pathol. Soc.*, 29, 425, 1878.
10. Lancefield, R. C., A serological differentiation of human and other groups of hemolytic streptococci, *J. Exp. Med.*, 57, 571, 1933.
11. Sherman, J. M., *Streptococcus lactis* and the streptococci of the so-called lactic group, *J. Dairy Sci.*, 38, 1184, 1955.
12. Jarvis, A. W. and Jarvis, D. D. W., Deoxyribonucleic acid homology among lactic streptococci, *Appl. Environ. Microbiol.*, 41, 77, 1981.
13. Garvie, E. I. and Farrow, J. A. E., *Streptococcus lactis* subspecies *cremoris* (Orla-Jensen) comb. nov. and *Streptococcus lactis* subspecies *diacetylactis* (Matuszewski et al.) nom. rev., com. nov., *Int. J. Syst. Bacteriol.*, 32, 453, 1983.
14. Niven, C. F., Smiley, K. L., and Sherman, J. M., The hydrolysis of arginine by streptococci, *J. Bacteriol.*, 61, 243, 1942.
15. Sandine, W. E., Elliker, P. R., and Hayes, H., Cultural studies on *Streptococcus diacetylactis* and other members of the lactic streptococcus group, *Can. J. Microbiol.*, 8, 161, 1962.
16. Pack, M. Y., Sandine, W. E., Elliker, P. R., Day, E. A., and Lindsay, R. C., Owades and Jakovac method for diacetyl determination in mixed strain starters, *J. Dairy Sci.*, 47, 981, 1964.
17. Hirsch, A. and Grimsted, E., The differentiation of the lactic streptococci and their antibiotics, *J. Dairy Res.*, 18, 198, 1951.
18. Cato, E. P. and Moore, W. E. C., A routine determination of the optically active isomers of lactic acid for bacterial classification, *Can. J. Microbiol.*, 11, 319, 1965.
19. Shattock, F. P. M., The streptococci of group D. The serological grouping of *Streptococcus bovis* and observations on serologically refractory group D strains, *J. Gen. Microbiol.*, 3, 80, 1949.
20. Hammer, B. W. and Baker, M. P., Classification of the *Streptococcus lactis* group, *Iowa Agric. Exp. Stn. Bull.*, 99, 1926.
21. Hucker, G. J. and Pederson, C. S., Studies on the coccaceae. XVI. The genus *Leuconostoc*, *N.Y. Agric. Exp. Stn. Bull.*, 142, 1930.
22. Matuszewski, T., Pijanowski, E., and Supinska, J., *Streptococcus diacetylactis* u. sp. i. jego zastosowanie przy wryobie masla., *Pol. Agric. Forest Annu.*, 36, 1, 1936.
23. Nelson, G. A. and Thornton, H. R., The lactic streptococci in Edmonton milks and creams, *Can. J. Technol.*, 30, 130, 1952.
24. Hirsch, A., Wild strains of *Streptococcus cremoris*, *J. Dairy Res.*, 19, 288, 1952.
25. Radich, P. C., An Ecological Study of the Lactic Streptococci, M.S. thesis, Oregon State University, Corvallis, 1968, 51.
26. Vedamuthu, E. R., Sandine, W. E., and Elliker, P, R., Flavor and texture in Cheddar cheese. I. Role of mixed strain lactic starter cultures, *J. Dairy Sci.*, 49, 144, 1966.
27. Gilliland, S. E., Sandine, W. E., and Vedamuthu, E. R., Acid-producing microorganisms, in *Compendium of Methods for the Microbiological Examination of Foods*, Speck, M. L., Ed., American Public Health Association, Washington, D.C., 1983, chap. 15.
28. Kempler, G. M. and McKay, L. L., Improved medium for the detection of citrate-fermenting *Streptococcus lactis* subsp. *diacetylactis*, *Appl. Environ. Microbiol.*, 39, 926, 1980.
29. Reddy, M. S., Vedamuthu, E. R., and Reinbold, G. W., A differential broth for separating lactic streptococci, *J. Milk Food Technol.*, 34, 43, 1971.
30. Reddy, M. S., Vedamuthu, E. R., Washam, C. J., and Reinbold, G. W., Agar medium for differential enumeration of lactic streptococci, *Appl. Microbiol.*, 24, 947, 1972.
31. Nickels, V. E. and Leesement, H., Methode zur Differenzierung and Quantitativen Bestimmung von Saureweckerbacterien, *Milchwissenschaft*, 19, 374, 1964.
32. Pack, M. Y., Vedamuthu, E. R., Sandine, W. E., Elliker, P. R., and Leesement, H., Effect of temperature on growth and diacetyl production by aroma bacteria in single and mixed strain lactic cultures, *J. Dairy Sci.*, 51, 339, 1968.
33. Terzaghi, B. E. and Sandine, W. E., Improved medium for lactic streptococci and their bacteriophages, *Appl. Microbiol.*, 9, 807, 1975.
34. Huggins, A. R. and Sandine, W. E., Differentiation of fast and slow milk-coagulating isolates in strains of lactic streptococci, *J. Dairy Sci.*, in press.

35. Lawrence, R. C., Thomas, T. D., and Terzaghi, B. E., Reviews in the progress of dairy science: cheese starters, *J. Dairy Res.*, 43, 141, 1976.
36. Cogan, T. M., Some aspects of the metabolism of dairy starter cultures, *Ir. J. Food. Sci. Technol.*, 7, 1, 1982.
37. Cogan, T. M., Citrate utilization in milk by *Leuconostoc cremoris* and *Streptococcus diacetylactics*, *J. Dairy Res.*, 42, 139, 1975.
38. Cogan, T. M., O'Dowd, M., and Mellerick, D., Effects of pH and sugar on acetoin production from citrate by *Leuconostoc lactis*, *Appl. Environ. Microbiol.*, 41, 1, 1981.
39. Drinan, E. F., Tobin, S., and Cogan, T. M., Citric acid metabolism in hetero- and homofermentative lactic acid bacteria, *Appl. Environ. Microbiol.*, 31, 481, 1976.
40. Harvey, R. J. and Collins, E. B., Citrate transport system of *Streptococcus diacetilactis*, *J. Bacteriol.*, 83, 1005, 1961.
41. Collins, E. B. and Harvey, R. J., Failure in the production of citrate permease by *Streptococcus diacetilactis*, *J. Dairy Sci.*, 45, 32, 1962.
42. McKay, L. L., Sandine, W. E., and Elliker, P. R., Lactose utilization by lactic acid bacteria: a review, *Dairy Sci. Abstr.*, 33, 493, 1971.
43. McKay, L. L., Miller, A., Sandine, W. E., and Elliker, P. R., Mechanisms of lactose utilization by lactic acid streptococci; enzymatic and genetic analyses, *J. Bacteriol.*, 102, 804, 1970.
44. Bisset, D. L. and Anderson, R. L., Lactose and D-galactose metabolism in group N streptococci: presence of enzymes for both the D-galactose-1-phosphate and D-tagatose-6-phosphate pathways, *J. Bacteriol.*, 117, 318, 1974.
45. Kandler, O., Carbohydrate metabolism in lactic acid bacteria, *Antonie van Leeuwenhoek; J. Microbiol. Serol.*, 49, 209, 1983.
46. Thomas, T. D., Heterolactic fermentations by homolactic starter streptococci resulting from growth with limiting carbohydrate, *N.Z. J. Dairy Sci. Technol.*, 14, 12, 1979.
47. Watson, J. D. and Crick, F. H. C., Molecular structure of nucleic acid. A structure for deoxribose nucleic acid, *Nature (London)*, 171, 737, 1953.
48. Davies, F. L., Underwood, H. M., and Gasson, M. J., The value of plasmid profiles for strain identification in lactic streptococci and the relationship between *Streptococcus lactis* 712, ML3, and C2, *J. Appl. Bacteriol.*, 51, 325, 1981.
49. Klaenhammer, T. R. and Sutherland, S. M., Detection of plasmid deoxyribonucleic acid in an isolate of *Lactobacillus acidophilus*, *Appl. Environ. Microbiol.*, 39, 671, 1980.
50. McKay, L. L., Functional properties of plasmids in lactic streptococci, *Antonie van Leeuwenhoek; J. Microbiol. Serol.*, 49, 259, 1983.
51. McKay, L. L. and Baldwin, K. A., Plasmid distribution and evidence for a proteinase plasmid in *Streptococcus lactis* C2, *Appl. Microbiol.*, 29, 546, 1975.
52. McKay, L. L., Baldwin, K. A., and Efstathiou, J. E., Transductional evidence for plasmid linkage of lactose metabolism, in *Streptococcus lactis* C2, *Appl. Environ. Microbiol.*, 32, 45, 1976.
53. Anderson, D. G. and McKay, L. L., Simple and rapid method for isolating large plasmid DNA from lactic streptococci, *Appl. Environ. Microbiol.*, 46, 549, 1983.
54. Orberg, P. K. and Sandine, W. E., A micro-scale method for rapid isolation of covalently closed circular plasmid DNA from group N streptococci, *Appl. Environ. Microbiol.*, 47, 677, 1984.
55. Collins, E. B., Action of bacteriophage on mixed strain cultures. V. Similarities among strains of lactic streptococci in commercially used cultures and use of a whey activity test for culture selection and rotation, *Appl. Microbiol.*, 3, 145, 1955.
56. Henning, D. R., Black, C. H., Sandine, W. E., and Elliker, P. R., Host-range studies of lactic streptococcal bacteriophages, *J. Dairy Sci.*, 51, 16, 1968.
57. Kempler, G. M. and McKay, L. L., Genetic evidence for plasmid linked lactose metabolism in *Streptococcus lactis* subspecies *diacetylactis*, *Appl. Environ. Microbiol.*, 37, 1041, 1979.
58. Kuhl, S. A., Larson, L. D., and McKay, L. L., Plasmid profiles of lactose negative and proteinase deficient mutants of *Streptococcus lactis* C10, ML3 and M18, *Appl. Environ. Microbiol.*, 37, 1193, 1979.
59. Kempler, G. M. and McKay, L. L., Characterization of plasmid deoxyribonucleic acid in *Streptococcus lactis* subspecies *diacetylactis:* evidence for plasmid linked citrate utilization, *Appl. Environ. Microbiol.*, 37, 316, 1979.
60. Efstathiou, J. D. and McKay, L. L., Inorganic salt resistance associated with a lactose-fermenting plasmid in *Streptococcus lactis*, *J. Bacteriol.*, 130, 257, 1977.
61. Hart, J. R., Davies, F. L., and Gasson, M. J., Plasmid control of nisin synthesis and sucrose fermentation in *S. lactis*, Abstr. paper presented at Lactic Acid Bacteria in Foods Symp., Wageningen, The Netherlands, September 1983.
62. Wulf, J. J., Lactic Streptococci: Growth Characteristics and Plasmid Profiles of Drug- and Bacteriophage-Insensitive Mutants, M.S. thesis, Oregon State University, Corvallis, 1981.

63. Allen, L. K., Sandine, W. E., and Elliker, P. R., Transduction in *Streptococcus lactis*, *J. Dairy Res.*, 30, 351, 1963.
64. Sandine, W. E. and Ayres, A. W., Method and Starter Compositions for Growth of Acid-Producing Bacteria (Insolubilized), U.S. Patent 4,382,965, 1983.
65. McKay, L. L., Baldwin, K. A., and Walsh, P. M., Conjugal transfer of genetic information in group N streptococci, *Appl. Environ. Microbiol.*, 40, 84, 1980.
66. Snook, R. J. and McKay, L. L., Conjugal transfer of lactose fermenting ability among *Streptococcus cremoris* and *Streptococcus lactis* strains, *Appl. Environ. Microbiol.*, 42, 904, 1981.
67. Gasson, M. J., Production, regeneration and fusion of protoplasts in lactic streptococci, *FEMS Microbiol. Lett.*, 9, 99, 1980.
68. Kondo, J. K. and McKay, L. L., Transformation of *Streptococcus lactis* protoplasts by plasmid DNA, *Appl. Environ. Microbiol.*, 43, 1213, 1982.
69. Chassy, B. M., Lee-Wickner, L. J., and Porter, E. V., Molecular characterization of plasmids and selected chromosomal genes isolated from *Lactobacillus* species., Abstr. paper presented at Lactic Acid Bacteria in Foods Symp., Wageningen, The Netherlands, September 1983.
70. Gibson, E. M., Chance, N. M., London, S. B., and London, J., Transfer of plasmid-mediated antibiotic resistance from streptococci to lactobacilli, *J. Bacteriol.*, 137, 614, 1979.
71. Pearce, L. E., Sage, M. F., and Hastings, J., Non-acid milk. The reappearance of an old problem, *N.Z. J. Milk Food Technol.*, 8, 165, 1973.
72. Anderson, A. W. and Elliker, P. R., The nutritional requirements of lactic streptococci isolated from starter cultures. II. A stimulatory factor required for rapid growth of some strains in reconstituted nonfat milk solids, *J. Dairy Sci.*, 36, 608, 1953.
73. Brown, R. J. and Swaisgood, H. E., Detection of Antibiotics in Milk, U.S. Patent 4,347,312, 1982.
74. Charm, S., Rapid Assay for Penicillin in Milk, monog. distributed by Penicillin Assays, Inc., Boston, 1978.
75. Cogan, T. M., Susceptibility of cheese and yogurt starters to antibiotics, *Appl. Microbiol.*, 23, 960, 1972.
76. Cogan, T. M. and Fitzgerald, G. F., A rapid method for the detection of penicillin in milk, *Ir. J. Food Sci. Technol.*, 4, 79, 1980.
77. Collins, E. B. and Tillion, A. W., Detection of antagonistic and symbiotic relationships among bacteria by diffusion chamber procedure, *J. Dairy Sci.*, 60, 387, 1977.
78. Czulak, J. and Hammond, L. A., Compatibility amongst strains of lactic streptococci, *Aust. J. Dairy Technol.*, 9, 115, 1954.
79. Lightbody, L. G. and Meanwell, L. G., The growth of lactic acid streptococci in mixed starter cultures, *Appl. Microbiol.*, 18, 53, 1955.
80. Nichols, A. A. and Ineson, P. J., Cheese starter recovery after attack by bacteriophage (strain dominance in multiple strain starters), *J. Dairy Res.*, 15, 99, 1947.
81. Vedamuthu, E. R., Hauser, B. A., Henning, D. R., Sandine, W. E., Elliker, P. R., and Day, E. A., Competitive growth of *Streptococcus diacetylactis* in mixed strain lactic cultures and cheese, *Proc. 17th Int. Dairy Congr.*, D2, 611, 1966.
82. Wulf, J. J., Thunell, R. K., and Sandine, W. E., Examination of strain interaction in a multiple strain lactic starter culture using streptomycin resistant mutants of lactic streptococci, *J. Dairy Sci.*, 66, 1436, 1983.
83. Leach, R. D. and Sandine, W. E., Numerical relationship between strains in frozen concentrates of lactic streptococcal starter cultures, *J . Dairy Sci.*, 59, 1392, 1976.
84. Thunell, R. K., Bodyfelt, F. W., and Sandine, W. E., Economic comparisons of Cheddar cheese manufactured with defined strain and commercial cultures, *J. Dairy Sci.*, 67, 1061, 1984.
85. Thunell, R. K., Sandine, W. E., and Bodyfelt, F. W., Phage insensitive multiple strain starter approach to Cheddar cheesemaking, *J. Dairy Sci.*, 64, 2270, 1981.
86. Thunell, R. K., Sandine, W. E., and Bodyfelt, F. W., Defined strains and phage insensitive mutants for commercial manufacture of cottage cheese and cultured buttermilk, *J. Dairy Sci.*, 67, 1175, 1984.
87. Collins, E. B., Influence of medium and temperature on endproducts and growth, *J. Dairy Sci.*, 60, 799, 1977.
88. Farr, S. M., Milk Fermenting Product and Method for Making Same, U.S. Patent 3,420,742, 1969.
89. Richardson, G. H., Cheng, C. T., and Young, R., Lactic bulk culture system utilizing a whey-based bacteriophage inhibitory medium and pH control. I. Applicability to American style cheese, *J. Dairy Sci.*, 60, 378, 1977.
90. Willrett, D. L., Sandine, W. E., and Ayres, J. W., Evaluation of a new pH-controlled bulk starter medium, *The Cheese Reporter*, 103, 8, 1979.
91. Sandine, W. E. and Ayres, J. W., Method and Starter Compositions for Growth of Acid-Producing Bacteria and Bacterial Compositions Produced Thereby, U.S. Patent 4,282,255, 1981.
92. Osborne, R. J. W. and Mabb, J. R., Production of frozen concentrated cheese starters by diffusion culture, *Proc. 19th Int. Dairy Congr.*, 1E(B6), 416, 1974.

93. Lindgren, B., Johnsson, H., and Leesement, H., Milk with reduced lactose content for preparation of lactic starter culture concentrates, *Proc. 19th Int. Dairy Congr.*, 1E(B6), 431, 1974.
94. Harvey, R. J., Damage, to *Streptococcus lactis* resulting from growth at low pH, *J. Bacteriol.*, 99, 1330, 1965.
95. Willrett, D. L., Sandine, W. E., and Ayres, J. W., Evaluation of pH controlled starter media including a new product for Italian and Swiss type cheeses, *Cult. Dairy Prod. J.*, 17, 5, 1982.
96. deMann, J. C., Rogosa, M., and Sharpe, E. M., A medium for the cultivation of lactobacilli, *J. Appl. Bacteriol.*, 23, 130, 1960.
97. Elliker, P. R., Anderson, A. W., and Hannesson, G., An agar culture medium for lactic acid streptococci and lactobacilli, *J. Dairy Sci.*, 39, 1611, 1956.
98. Elliker, P. R., "This week's citation classic," *Curr. Cont.*, 14, 22, 1983.
99. Citti, J. E., Sandine, W. E., and Elliker, P. R., Comparison of slow and fast acid producing *Streptococcus lactis, J. Dairy Sci.*, 48, 14, 1965.
100. Pearce, L. E., Skipper, N. A., and Jarvis, B. D. W., Proteinase activity in slow lactic acid-producing variants of *Streptococcus lactis, Appl. Microbiol.*, 27, 933, 1974.
101. Limsowtin, G. K. Y. and Terzaghi, B. E., Agar medium for the differentiation of fast and slow coagulating cells in lactic streptococcal cultures, *N.Z. J. Dairy Sci. Technol.*, 11, 65, 1976.
102. Huggins, A. R. and Sandine, W. E., Selection and characterization of phage insensitive lactic streptococci, *J. Dairy Sci.*, 62, 70, 1979.
103. Willrett, D. L., Development of Internally pH-Controlled Bulk Starter Media for the Propagation of Lactic Acid Bacteria, Ph.D. thesis, Oregon State University, Corvallis, 1982.
104. Christensen, V. W., Recent developments in starter techniques, *Dairy Ind.*, 37, 655, 1972.
105. Gulstrom, T. J., Pearce, L. E., Sandine, W. E., and Elliker, P. R., Evaluation of commercial phage inhibitory media, *J. Dairy Sci.*, 62, 208, 1979.
106. Hull, R. R., Control of bacteriophages, in cheese factories, *Aust. J. Dairy Technol.*, 32, 65, 1977.
107. Lawrence, R. C. and Pearce, L. E., Cheese starters under control, *Dairy Ind.*, 37(2), 73, 1972.
108. Sandine, W. E., Lactic starter culture technology, Pfizer Inc., New York, *Pfizer Cheese Monogr.*, 7, 55, 1979.
109. Czulak, J., Bant, D. J., Blyth, S. C., and Cace, J., A new cheese starter system, *Dairy Ind. Int.*, 44, 17, 1979.
110. Marshall, R. J. and Berridge, N. J., Selection and some properties of phage-resistant starters for cheese making, *J. Dairy Res.*, 43, 449, 1976.
111. Limsowtin, G. K. Y. and Terzaghi, B. E., Phage resistant mutants: their selection and use in cheese factories, *N.Z. J. Dairy Sci. Technol.*, 11, 251, 1976.
112. Richardson, G., Hong, G. L., and Ernstrom, C. A., Defined single strains of lactic streptococci in bulk culture for Cheddar and Monterey cheese manufacture, *J. Dairy Sci.*, 63, 1981, 1980.
113. Daly, C., The use of mesophilic cultures in the dairy industry, *Antonie van Leeuwenhoek; J. Microbiol. Serol.*, 49, 297, 1983.
114. Teuber, M. and Lembke, J., The bacteriophages of lactic acid bacteria with emphasis on genetic aspects of group N lactic streptococci, *Antonie van Leeuwenhoek; J. Microbiol. Serol.*, 49, 283, 1983.
115. Forsin, R. and Häivä, V.-M., Induction of stable slime-forming and mucoid states by pfluorophenylalanine in lactic streptococci, *FEMS Microbiol. Lett.*, 12, 409, 1981.
116. Thunell, R. K., Sandine, W. E., and Bodyfelt, F. W., Frozen starters from internal pH-control-grown cultures, *J. Dairy Sci.*, 67, 24, 1984.
117. Hunter, G. J. E., Examples of variations in pure cultures of *Streptococcus cremoris, J. Dairy Res.*, 10, 464, 1939.
118. Limsowtin, J. K. Y., Heap, H. A., and Lawrence, R. C., Heterogeneity among strains of lactic streptococci, *N.Z. J. Dairy Sci. Technol.*, 13, 1, 1978.
119. Salih, M. A. and Sandine, W. E., A rapid test for detecting lactic streptococcal agglutinins in cheese milk, *J. Dairy Sci.*, 67, 7, 1984.

Chapter 3

THE LEUCONOSTOCS: MILK PRODUCTS

Timothy M. Cogan

TABLE OF CONTENTS

I. HISTORICAL

The use of starters to initiate the fermentation of milk and cream in the manufacture of cheese and butter had been practiced long before anything was known about the bacteria they contained or indeed before it was appreciated that bacteria were involved at all. Initially, the milk or cream was allowed to stand at room temperature and the lactic acid bacteria adventitiously present multiplied and coagulated the milk or cream. These haphazard methods were used until the end of the 19th century when Conn in the U.S. and Storch in Denmark showed that a good flavored butter "free of souring defects" was produced when the cream was soured with pure cultures of *Streptococcus lactis*. The truly fine flavor of the traditional butter was still missing. The reasons for this became apparent in 1919 when three independent groups of workers (Boekhout and Ott de Vries[1] in Holland, Storch[2] in Denmark, and Hammer and Bailey[3] in the U.S.) established that lactic cultures that showed good flavor development were in fact a mixture of two different types of lactic acid bacteria, one of which was responsible for acid production (*S. lactis* and *S. cremoris*) and the other for flavor production. The acid producers are covered in other chapters of this book and will not be discussed further here. Initially, the flavor producers were shown to be *Leuconostoc* species, but in 1936 additional flavor organisms were isolated independently by Matusewski et al.[4] in Poland, who called them *S. diacetilactis* and by van Beynum and Pette[5] in Holland, who called them *S. citrophilus*. They were subsequently shown to be similar[6] and in the most recent edition of *Bergey's Manual of Determinative Bacteriology*[7] they have been given the name of *S. lactis* subsp. *diacetylactis*. The only difference between this organism and *S. lactis* is its ability to dissimilate citrate to a variety of end products so that this organism functions in both acid and flavor production.

This chapter is concerned with the *Leuconostoc* species, but reference will be made also to *S. lactis* subsp. *diacetylactis* as it would be impossible to review the properties conferred by the leuconostocs on lactic starter cultures without also considering the somewhat different properties conferred by *S. lactis* subsp. *diacetylactis*.

II. FUNCTIONS OF THE FLAVOR PRODUCERS

The flavor-producing organisms have two major functions, one of which is rather obvious, viz. the production of flavor compounds (e.g., diacetyl and acetate) in unripened dairy products such as cultured buttermilk, sour cream, lactic butter, etc. They are also responsible for eye formation through CO_2 production in many cheese varieties, e.g., Edam, Gouda, Tilsit, etc. Both functions are due to the metabolism of citrate but *Leuconostoc* species also produce CO_2 from lactose. Leuconostocs and lactobacilli have been incriminated as the cause of open texture (slit openness) in Cheddar cheese,[8,9] leuconostocs as the cause of early gas in Gouda cheese,[10] and *S. lactis* subsp. *diacetylactis* as the cause of floating curd in cottage cheese.[11] All of these defects are due to production of CO_2 from citrate and in some cases perhaps also from lactose. Despite this, mixed strain starters containing leuconostocs and *S. lactis* subsp. *diacetylactis* are still used commercially in cheese manufacture[12] so that there must be other contributory factors to these defects. In this regard it is noteworthy that Galesloot[10] states that the *Leuconostoc* species normally present in starters are not dangerous, but only those that grow rapidly in *plain* (i.e., unfortified) milk. Generally *Leuconostocs* grow very poorly in milk. To prevent floating curd, cottage cheese is now generally made with a culture containing no flavor producers. However, these may be added to the curd after manufacture as a creaming mixture consisting of either a citrated milk culture of *L. cremoris*,[13] a citrated whey culture of *S. lactis* subsp. *diacetylactis*,[14] or a concentrated culture of *L. cremoris*.[15]

III. TYPES OF CULTURES

Three different types of starter cultures are used in the dairy industry: single, multiple, and mixed strain cultures. These are rigorously treated in other chapters of this book and will not be further reviewed here except for a few comments. Multiple strain cultures are mixtures of known, compatible, nonphage related, carefully selected strains while mixed cultures are stable mixtures of unknown strains that have evolved through continual use and been found to give generally consistent products when used commercially. The number of strains and their individual phage relationships in these mixed cultures is not known though their overall phage relationships may be.

Depending on the nature of the flavor producers, mixed strain starter cultures are usefully divided into:[16]

B cultures: *Leuconostoc* species as the sole flavor producer
D cultures: *S. lactis* subsp. *diacetylactis* as the sole flavor producer
BD cultures: Both *Leuconostoc* species and *S. lactis* subsp. *diacetylactis* as flavor producers
O or N cultures: No flavor producers

The sobriquet "B" stems from the first letter of the old genus name for *Leuconostoc* viz. *Betacoccus* while the "D" stems from the first letter of *diacetylactis*, so that these cultures would be more correctly called L (from *Leuconostoc*), S (from *Streptococcus*), LS and O cultures, respectively, despite the fact that O cultures also contain streptococci. Nevertheless the old naming system is still widely used and will be used in this review. To complicate things further, the flavor producers are often generically called aroma producers, citrate utilizers, or citrate fermenters. The use of the adjective "citrate" is easily understood since citrate is the precursor of the flavor compounds, but to call them citrate fermenters is incorrect since it implies that they can use citrate as an energy source which is not true.

IV. TAXONOMY

Despite the fact that leuconostocs have been known components of starter cultures since the 1920s, factual information on the species of leuconostocs involved is difficult to find in the literature. There are several reasons for this, notably the difficulty in deciding their taxonomic status. Many of the early workers were more interested in what determined flavor formation than in bacterial taxonomy, and in fact two of the most important workers (Orla-Jensen and Hammer) did not separate *Pediococci* from *Leuconostocs*.[17] The definitive study of their taxonomy is that of Garvie[17] who divided the genus into four species: *L. cremoris* (*citrovorum*), *L. lactis*, *L. dextranicum*, and *L. mesenteroides*. Two further species, *L. paramesenteroides* and *L. oenos*, have since been added.[18,19] *L. oenos* is completely different from the other species in that it is acidogenic (growing at a pH of 4.8), is found only in wine, and many strains require a growth factor present in tomato juice for growth.[19] The five nonacidogenic species (*L. cremoris*, *L. lactis*, *L. dextranicum*, *L. mesenteroides*, and *L. paramesenteroides*) are phenotypically homogeneous. This is in keeping with their separation on the basis of differences in fermented sugars since such differences reflect the presence or absence of a few enzymes (and hence genes) involved in sugar transport or the early portion of sugar metabolism. The fifth nonacidogenic species (*L. cremoris*) is characterized by its ability to ferment only those sugars associated with milk, lactose, galactose, and glucose. This may be a simple evolutionary response since the only known source of this species is dairy cultures. DNA-DNA hybridization studies have suggested four species,

Table 1
USEFUL CRITERIA FOR DISTINGUISHING THE FLAVOR-PRODUCING MICROORGANISMS IN LACTIC CULTURES

Organism	Sugar fermentation[a,24]	Lactic acid[b]	Citrate utilization[25]	Acetoin prod. from citrate[26]	Growth at 37°C	Growth in litmus milk[e]	Dextran prod.
L. cremoris	Hetero	D	+	–[c]	–	–	–
L. mesenteroides	Hetero	D	+	–[c]	+[d]	–	+
L. lactis	Hetero	D	+	–[c]	+	A	–
S. lactis subsp. *diacetylactis*	Homo	L	+	+	+	RA ± C	–

[a] The superscript number refers to the reference for the method commonly used for the determination.
[b] Commercial kits are available for the determination of the isomer of lactic acid produced.
[c] Most strains will produce acetoin from citrate at low pH.
[d] >90% of strains are positive.
[e] A, acid; R, reduction; C, coagulation.

L. oenos, L. lactis, L. paramesenteroides, and *L. mesenteroides,* with *L. cremoris* and *L. dextranicum* being subspecies of *L. mesenteroides.*[20,21]

With this background it is difficult to decide which species of leuconostoc is found in dairy starter cultures. A study of 45 leuconostocs isolated from starter cultures for sour cream by Schlegel and Babel[22] showed that 47% were *L. cremoris* (*citrovorum*) and 34% *L. mesenteroides;* the remaining 19% differed from any recognized species of leuconostoc. Galesloot and Hassing[16] and Garvie[20] feel that *L. cremoris* is the main *Leuconostoc* species while Stadhouders[23] stated that both *L. lactis* and *L. cremoris* are found. Galesloot and Hassing[16] also refer to work done on Dutch cultures in 1936 and 1948 that showed that *Leuconostoc* species (generally *L. cremoris*) were the sole flavor producers. Dutch starters examined in 1954 contained both *L. cremoris* and *S. lactis* subsp. *diacetylactis* as flavor producers (so-called BD cultures). A study in 1976 of several mixed cultures used commercially in the U.K. and Ireland revealed that BD cultures were the most prevalent.[12] A simple scheme for distinguishing the flavor producers found in mixed starter cultures is shown in Table 1.

Except for *L. lactis,* most *Leuconostoc* species grow poorly in milk unless a stimulant like 0.3% (w/v) yeast extract is added and in a few cases 1.0% (w/v) glucose.[17] The latter reflects an inability of some strains to use lactose as an energy source while the former probably reflects their inability to produce sufficient proteinase to hydrolyze milk proteins to the amino acids and small peptides required for growth. All leuconostocs require valine and glutamate for growth, most require alanine, and most also are stimulated by methionine.[18] They show considerable variation in their requirements for the remaining amino acids with many strains requiring eight or more.[18] Lawrence et al.[27] have calculated that milk contains only 5 to 20% of the concentrations of free amino acids and peptides required for maximum growth of the lactic streptococci and presumably also the leuconostocs. Yeast extract is considered a rich source of amino acids and peptides and this would appear to be its function in stimulating growth of the leuconostocs.

Despite their poor growth in milk, there are reports of *Leuconostoc* species attaining counts of 10^8 to 10^9 cells per milliliter in unfortified milk with an average generation time of 3.6 ± 0.89 hr (for 9 strains) at 22°C.[28,29] Counts of this magnitude should show some reaction in milk since a coagulated culture is generally assumed to contain 10^9 cells per milliliter. Such data should be interpreted with caution because of the difficulty in deciding whether an organism is a leuconostoc or not. However, two of the strains produced fairly high levels of ethanol (400 to 500 μg/mℓ) in the unfortified milk and presumably were heterofermentative.[28]

In complex broth media, most leuconostocs grow well on overnight incubation at 30°C. *L. cremoris* is an exception with many strains requiring 2 or 3 days incubation before much turbidity due to growth is seen (unpublished data). The addition of cysteine to broth media promotes growth of *L. cremoris*.[18]

V. ISOLATION AND ENUMERATION

Leuconostocs generally withstand acid conditions much better than lactic streptococci so that an overincubated mixed culture is a useful starting point for their isolation. Numerous media have been described for their enumeration. The most useful of these have been compared by Waes.[30] The most difficult enumeration is an individual count of leuconostocs and *S. lactis* subsp. *diacetylactis* in a BD culture. Galesloot et al.[31] developed three partially selective and differential media for enumerating leuconostocs and *S. lactis* subsp. *diacetylactis* in B, D, and BD cultures which are useful but complicated to use. These were subsequently modified to one medium (called KCA) which currently is the medium of choice.[32] All of these media are opaque because of the presence of insoluble calcium citrate. Utilization of the citrate by the *Leuconostoc* species and *S. lactis* subsp. *diacetylactis* produces clear haloes around the colonies, but these must be distinguished by further testing. The opacity of the medium does not allow one to count the noncitrate utilizing bacteria in mixed cultures (i.e., *S. lactis* and *S. cremoris*), but this can be accomplished by addition of 0.01% (w/v) triphenyltetrazolium chloride (TTC) (unpublished data). All bacteria in the mixed culture reduce the soluble and colorless TTC to an insoluble red complex that precipitates on each colony during growth, thus making them visible. The citrate utilizing bacteria in addition are surrounded by a clear halo as a result of metabolizing the insoluble calcium citrate. Suitable dilutions (0.1 mℓ aliquots) of the starter are spread on the surface of predried plates which are incubated anaerobically at 30°C for 48 to 72 hr. A count of all red colonies gives the total count while a count of red colonies surrounded by a clear halo gives the number of citrate utilizing bacteria (both leuconostocs and *S. lactis* subsp. *diacetylactis*). With experience it is fairly easy to distinguish *S. lactis* subsp. *diacetylactis* from leuconostocs as the latter give smaller colonies surrounded by smaller zones of clearing. However, they can be routinely distinguished by inoculating each colony that produces a clear zone into milk or a complex broth containing both citrate and a fermentable sugar (e.g., MRS,[33] which has about the same citrate level as milk) and incubating at 30°C until turbid growth is obtained (about 16 to 30 hr). The cultures are then checked for citrate utilization and acetoin production by the procedures of Marier and Boulet[25] and Westerfeld,[26] respectively. Leuconostocs utilize citrate but do not produce acetoin while *S. lactis* subsp. *diacetylactis* utilizes citrate and produces acetoin (Table 1). (Studies in our laboratory have shown that both tests are easily quantified in milk or broth. However, the standard curve for acetoin must be determined in the uninoculated medium since the medium decreases the sensitivity of the method to detect acetoin.) The reason for rechecking citrate utilization is that occasionally zone producers are isolated which do not utilize citrate; the zones are due to solubilization of the calcium citrate by high levels of lactate produced by *S. lactis* or *S. cremoris*. If citrate utilization were not checked, these streptococci would be incorrectly identified as leuconostocs.

Nonselective differential agar media for distinguishing between the lactic streptococci (*S. lactis*, *S. cremoris*, and *S. lactis* subsp. *diacetylactis*) have been proposed by Reddy et al.[34] and Kempler and McKay.[35] The behavior of *Leuconostoc* species on

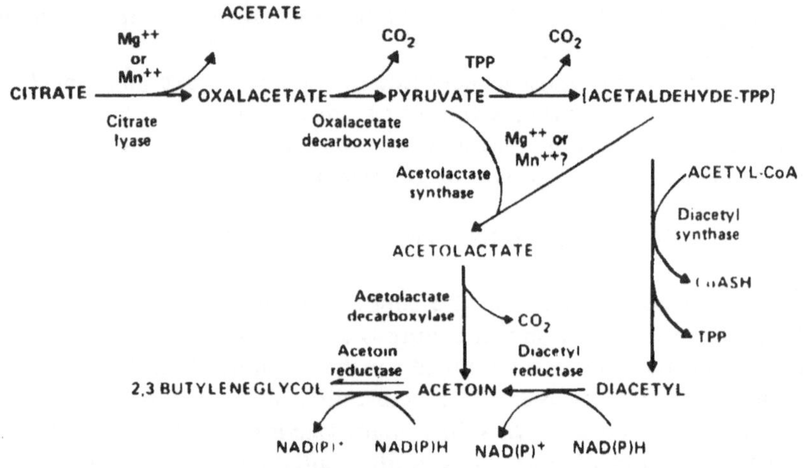

FIGURE 1. Pathway for citrate metabolism in *Leuconostoc* species and in *Streptococcus lactis* subsp. *diacetylactis.*

such media has not been reported but we (unpublished data) have found that Kempler and McKay's medium is not suitable for leuconostocs since many grow poorly and many citrate-utilizing strains fail to indicate citrate utilization.

None of these media are suitable for isolating *S. lactis* subsp. *diacetylactis* or leuconostocs from cheese, since lactobacilli, which form a considerable portion of the microflora of ripening cheese, can also grow and most of them utilize citrate.[36] Under such conditions, each isolate would have to be examined microscopically, tested for the sugar fermentation and its ability to produce acetoin and metabolize citrate. Citrate-metabolizing heterofermentative lactobacilli, e.g., leuconostocs, do not produce acetoin while citrate-utilizing homofermentative lactobacilli, such as *S. lactis* subsp. *diacetylactis*, do produce acetoin.[37]

VI. CITRATE METABOLISM

Most of the research on citrate metabolism in lactic acid bacteria has been carried out with *S. lactis* subsp. *diacetylactis*, but there is no reason to believe that citrate metabolism in *Leuconostoc* species would be any different except that acetoin and diacetyl are produced only at low pH. Neither organism can use citrate as an energy source though they will metabolize it rapidly in the presence of a metabolizable energy source. The pathway is shown in Figure 1. Citrate is transported into the cell by means of a pH-dependent inducible permease.[38] It is then converted to acetate and oxalacetate by citritase, which is constitutively present in *S. lactis* subsp. *diacetylactis*[39,40] and inducible in leuconostocs.[41] Oxalacetate is decarboxylated to CO_2 and pyruvate which is further decarboxylated to CO_2 and a TPP-acetaldehyde complex.[42] This complex can react with another molecule of pyruvate to form α-acetolactate[42] in a reaction catalyzed by acetolactate synthase which is constitutive in both *S. lactis* subsp. *diacetylactis* and leuconostocs.[40,41] Acetolactate is further decarboxylated to CO_2 and acetoin.[42] Acetoin may also be produced from diacetyl by diacetyl reductase (DR) in an NADH-requiring reaction (*S. lactis* subsp. *diacetylactis* and some leuconostocs)[40-43] or NADPH (some leuconostocs and heterofermentative lactobacilli).[41] This may be unlikely in vivo, at least in *S. lactis* subsp. *diacetylactis* because of its limited ability to produce acetyl CoA, one of the precursors of diacetyl.[44] Butylene glycol is produced from acetoin in another NADH/NADPH-requiring reaction catalyzed by acetoin reductase (AR). In

Enterobacter aerogenes the reduction of diacetyl and acetoin is carried out by the same enzyme,[45] but whether this is true in the lactic acid bacteria is unknown.

There has been some debate as to whether α-acetolactate is involved in diacetyl synthesis in bacteria (see Stadhouders[23] for further discussion). Acetolactate is an unstable compound and is easily decarboxylated (nonenzymatically and at low pH) to acetoin and diacetyl. Rates of formation of acetoin and diacetyl from 10 μM acetolactate of 0.43 μM/hr and 0.19 μM/hr, respectively, at pH 4.5 can be calculated from data reported by Collins and Speckman.[46] These rates were almost linear over an 8-hr period. Jönsson and Pettersson[47] have shown that low pH, low O-R potential, and distillation in an inert atmosphere decrease the oxidation of acetolactate to diacetyl but increase the proportion recovered as acetoin. They have also shown that B, BD, and D cultures produce acetolactate while Collins and Speckman[46] found that two strains of *S. lactis* subsp. *diacetylactis* did not. It is difficult to reconcile these contradictory reports. However, the levels of acetoin produced by the two strains of *S. lactis* subsp. *diacetylactis* in the latter report are very low compared to other reports,[37,40,49,50] suggesting that the strains may have lost the ability to metabolize citrate and hence their ability to produce acetolactate. Differences in methodology may also explain these contrasting results. Collins and Speckman[46] determined acetolactate after decarboxylating it with H_2SO_4 to acetoin, which was measured by the well-tried Westerfeld[26] procedure (the results were corrected for any acetoin produced by the organism), while Jönsson and Pettersson[48] used a little-tried method based on the ability of acetolactate to reduce ammonium molybdate to molybdene blue.

Pyruvate is also an intermediate in carbohydrate metabolism by both leuconostocs and streptococci, yet no diacetyl and little if any acetoin is produced from carbohydrates unless citrate or some other source of pyruvate is also present.[37,50] The stoichiometry of sugar fermentation demands that every mole of NADH formed must be reoxidized to NAD^+ to continue the fermentation. In the case of lactic acid bacteria this is accomplished through the production of lactic acid and ethanol in reactions catalyzed by lactic dehydrogenase and alcohol dehydrogenase, respectively:

$$\text{pyruvate} + \text{NADH} \rightarrow \text{lactate} + \text{NAD}^+$$
$$\text{acetaldehyde} + \text{NADH} \rightarrow \text{ethanol} + \text{NAD}^+$$

In the presence of citrate, pyruvate is produced without concomitant production of NADH. There is thus an excess of pyruvate over that necessary to reoxidize NADH. This excess is metabolized through the production of diacetyl and acetoin.[51]

VII. PRODUCTION OF DIACETYL, ACETOIN, AND 2,3-BUTYLENE GLYCOL

A. Pure Culture

Growing cultures of *S. lactis* subsp. *diacetylactis* and leuconostocs behave completely differently in the production of diacetyl and acetoin. As one would suspect, no diacetyl and only traces of acetoin are produced in the absence of citrate.[40,50] In media containing citrate *S. lactis* subsp. *diacetylactis* begins to produce diacetyl and acetoin as soon as growth begins and both compounds reach a maximum at the point of complete disappearance of citrate.[37,40,49,50] Diacetyl and acetoin accumulate because citrate represses the synthesis of DR and AR so that there is little or no enzyme to reduce them to acetoin and 2,3-butylene glycol, respectively.[40] Once citrate metabolism is finished, increased synthesis or DR and AR occurs resulting in decreased levels of diacetyl and acetoin.[37,49,50] Reduction of acetoin is not observed in all cases, so other factors also must be involved. Acetoin production is always much greater than that of diacetyl.

The ratio has been reported to be as high as 43:1 and the amount of citrate converted to both compounds ranges from 30 to 92%.[37,50] Growth of the strains that converted low levels was constrained for some reason, thus the results may not be representative. The low production of diacetyl is thought to be due to a limited ability to synthesize acetyl-CoA which is involved only in diacetyl biosynthesis.[44] On the other hand, growing cultures of *Leuconostoc* species metabolize citrate without producing diacetyl or acetoin at neutral pH[37,41,49,52,53] even though DR and AR are repressed by citrate;[41] however, acetate is produced.[53] Under acid conditions citrate is utilized and both diacetyl and acetoin are produced by many[37,39,52-56] but not all strains.[52] When produced by *Leuconostoc* species, acetoin also is quantitatively a more important metabolite than diacetyl.[37,49,52,54,56]

The effect of pH on acetoin (and diacetyl) production from citrate by *Leuconostocs* species has been known since the 1940s yet the mechanism(s) remains obscure. A detailed study of the effect of pH on citrate metabolism and acetoin production by cell suspensions of *L. lactis* NCWI (in the absence of an energy source) has recently been reported.[52] Citrate utilization and acetoin production were not linear throughout the incubation period. However, the rates of citrate utilization and acetoin production have been calculated using the data from the linear portion of the relevant curves. It is clear that citrate metabolism occurs at least over the pH range 6.5 to 4.3 and is optimum at pH 5.4. In contrast, acetoin production was nil in phosphate buffer at pH 6.0 or 6.5, but increased as the pH decreased from 6.5 in acetate buffer. No optimum was observed, but acetoin production reached a maximum at pH 4.3 where it was quantitatively recovered from citrate. Whether phosphate inhibited or acetate stimulated acetoin production was not determined.

Metabolizable sugars (> 10 m*M*) dramatically increased the rate of citrate uptake but totally inhibited acetoin formation at pH 5.4.[52] Reducing the pH to 4.5 reduced the inhibition somewhat but it was still quite marked. Further studies in our laboratory have revealed that this is due to inhibition of acetolactate synthase by many of the intermediates of the phosphoketolase pathway (6-phosphoglucic acid, 3-phosphoglyceric acid, 2-phosphoglyceric acid, phospheonol pyruvate, and ATP). Reducing the pH to 4.5 relieved the inhibition of many of these compounds. Lactate dehydrogenase (LDH) is also involved in the regulation of acetoin production in the leuconostocs. As the pH decreases, the kanamycin (Km) of LDH for pyruvate also decreases.[90] At acid pH, less pyruvate is required to saturate LDH and any excess would be available for other reactions involving pyruvate.

Definitive information on 2,3-butylene glycol production by the *Leuconostocs* and *S. lactis* subsp. *diacetylactis* is lacking in the literature. The main reasons for this is the difficulty in measuring it. Until the advent of gas-liquid chromatography (GLC), the method used relied on oxidation of the glycol to acetoin and it was not clear what effect this had on acetoin itself or if values reported were corrected for any acetoin that was present initially. If 2,3-butylene glycol formation occurs it is probably in fairly small amounts, because in many strains of *S. lactis* subsp. *diacetylactis* the recovery rate of acetoin from citrate is high,[50] and in addition, the AR levels in both *S. lactis* subsp. *diacetylactis* and *Leuconostoc* species are quite low.[40,41] Values for *L. cremoris* of 1.22 m*M* at pH 4.3 in milk containing 13.21 m*M* citric acid have been reported,[22] but recovery of the citrate as diacetyl, acetoin, and 2,3-butylene glycol was only 55% at this point.

B. Mixed Cultures

Mixed strain starters containing either *S. lactis* subsp. *diacetylactis* or *Leuconostoc* species as aroma producers behave differently with regard to the rates and extent of citrate metabolism in milk, and hence their rates of acetoin and diacetyl production

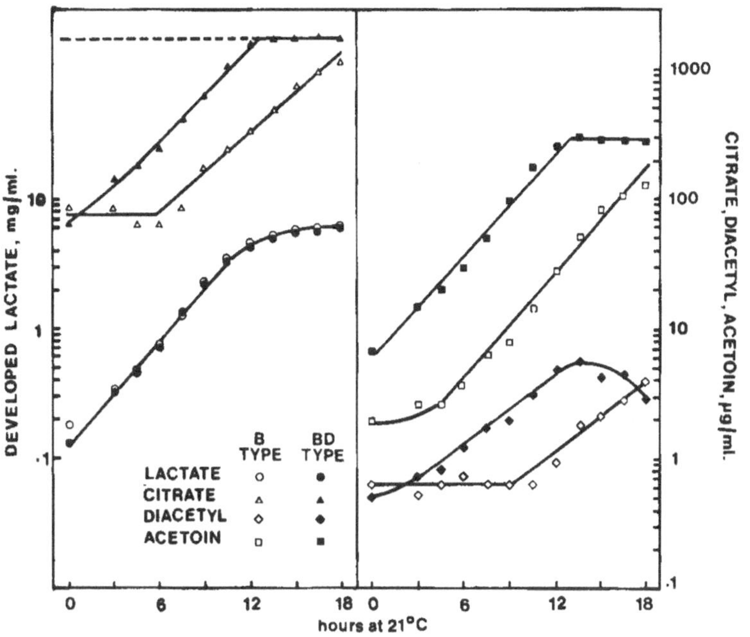

FIGURE 2. Citrate utilization and production of diacetyl, acetoin, and lactate by B and BD cultures in milk at 21°C.

differ. D cultures begin to metabolize citrate (and consequently produce acetoin and diacetyl) as soon as growth begins, while B cultures show a lag in citrate metabolism (and acetoin and diacetyl production). In fact, D cultures will generally metabolize all the citrate in milk within 10 hr at 21°C (1 to 2% inoculum) while some citrate is often still present in B cultures after 18 hr at 21°C (1 to 2% inoculum).[16,47,57] BD cultures behave like D cultures.[47,57] It has been suggested that this difference in rate of citrate metabolism is due to a higher number of flavor producers in D than in B starters,[16] but it is more likely to reflect more active growth and consequent metabolism by *S. lactis* subsp. *diacetylactis* compared to leuconostocs.

There is a common belief that little acetoin or diacetyl accumulate in mixed cultures until the pH drops to 5.0 or 5.5.[58,59] This may be true of B cultures because they show a lag before acetoin and diacetyl production become exponential — it is not true of BD and D cultures. Part of the reason for this is that time-course data were often plotted arithmetically. Such plots may not be quite correct since the low production levels associated with poor growth are not clearly seen. Semilogarithmic plotting of the data shows that citrate metabolism and diacetyl and acetoin production begin as soon as acid production (growth) begins.[47,57] Further, pure cultures of *S. lactis* subsp. *diacetylactis* also produce considerable amounts of diacetyl and acetoin before a pH of 5 is reached.[50]

Maximum production of acetoin and diacetyl occurs at the point of exhaustion of citrate from the milk. At this point D and BD cultures produce an average of 3.3 m*M* acetoin although one culture produced 5.6 m*M*, while B cultures reach a maximum of 1 to 1.4 m*M*.[47,57] In all three types of culture diacetyl production averages about 0.05 m*M*.[47,57] A comparison of citrate metabolism and diacetyl, acetoin, and lactic acid production by a B and a BD culture in milk at 21°C is shown in Figure 2. The rate and extent of lactic acid production is the same in both cultures. This is to be expected since there are no differences in the rate of lactic acid production between B, D, and BD cultures or indeed between pure strain and mixed strain cultures. At 21°C all have a generation time of 2.2 ± 0.2 hr.[60]

The question might be asked, how are diacetyl and acetoin produced in B cultures when neither compound is produced by pure cultures of *Leuconostoc* species? The main reason would appear to be that leuconostocs grow poorly in mixed culture. Because of this sufficient citrate remains in the milk to act as a precursor under acid conditions.[51]

VIII. CARBOHYDRATE METABOLISM

Both sugar transport and sugar metabolism are quite different in *S. lactis* subsp. *diacetylactis* and leuconostocs. *S. lactis* subsp. *diacetylactis* is probably like *S. lactis* and *S. cremoris* which transport sugar by the phosphoenolpyruvate (PEP) phosphotransferase system (PTS) in which sugar is phosphorylated by PEP in a complex series of reactions as it is transported across the cell membrane (for a review see Reference 61). The leuconostocs on the other hand do not use the PEP/PTS system for sugar transport; instead they use an ATP-energized system.[62] If they used the PEP/PTS system, all the PEP produced in the fermentation would be used to transport the sugar leaving none available for the production of the ATP necessary for other biosynthetic reactions.[62] Based on this argument uptake of disaccharides such as lactose (and other polysaccharides) would be theoretically possible since 2 mol PEP would be produced per mol lactose fermented, only one of which would be necessary for lactose transport. Whether this occurs in practice has not been studied.

Sugar fermentation by *S. lactis* subsp. *diacetylactis* and the other lactic streptococci, occurs via the glycolytic pathway in which 2 mol lactic acid are produced per mol hexose metabolized resulting in a net gain of 2 mol ATP. On the other hand, sugar fermentation by *Leuconostoc* species occurs via the phosphoketolase pathway in which 1 mol each of lactate, ethanol, and CO_2 are produced per mol hexose metabolized resulting in a net gain of 1 ATP. The regulation of sugar transport and metabolism in the streptococci has been thoroughly studied. In contrast, regulation of sugar transport and metabolism in the leuconostocs has not been studied at all.

IX. GENETICS

It is curious that the main industrially important parameters of lactic starter cultures, lactose, citrate, and protein metabolism are all encoded by plasmid DNA, which is easily lost in many strains. Once lost such strains cannot metabolize the compound for which that plasmid coded nor can they regain the ability to do so except by gene transfer.

A review of the genetics of starter bacteria is outside the ambit of this chapter and in any case has been thoroughly done for lactic acid bacteria by Davies and Gasson[69] and for *S. lactis* subsp. *diacetylactis* by Kempler and McKay.[70] Work on the genetics of the leuconostocs is extremely limited. As far as this author is aware, there is only one report dealing with it.[91] He found that citrate negative mutants of *L. mesenteroides* X2 that failed to transport citrate lacked an 11.0 Mdal plasmid and citratase, but contained acetolactate synthase, diacetyl reductase, and acetoin reductase. These data imply that the plasmid codes for citrate permease. Kempler and McKay[71] have reported that citrate permease activity in *S. lactis* subsp. *diacetylactis* is associated with a 5.5 Mdal plasmid. Lactose uptake and proteinase production are plasmid coded in the lactic streptococci. Whether this is also true of the leuconostocs has not been studied; however, O'Sullivan[91] has found some evidence for plasmid involvement in lactose metabolism.

X. FACTORS AFFECTING OPTIMUM PERFORMANCE

A. Flavor Compound Levels

From the point of view of flavor production, diacetyl is the most important metabolite; its reduced products, acetoin and 2,3-butylene glycol are flavorless. There is some evidence that acetate is also involved in flavor[72] and it is not inconceivable that CO_2 may be involved as well when one considers the dramatic effect that it has on the taste of carbonated drinks. Citrate is the precursor of the metabolites although CO_2 is also produced from sugar by the leuconostocs. Milk contains approximately 8 m*M* citrate so that the theoretical maxima are 4 m*M* (diacetyl + acetoin + 2,3-butylene glycol) (350 μg/mℓ calculated as acetoin), 8 m*M* acetate (360 μg/mℓ) and 16 m*M* CO_2 (700 μg/mℓ), much of which would be dissipated into the atmosphere since the solubility of CO_2 at 20°C is only 4 m*M*. In addition, the solubility of CO_2 is markedly decreased by lowering the pH and increasing temperature.

Definitive information on the exact level of diacetyl necessary to impart a good flavor to fermented dairy products appears to be unavailable in the literature. The optimum is generally considered to lie somewhere between 1.5 and 2.5 μg/g. The products in which diacetyl is important include sour cream, cultured buttermilk, lactic butter, and cottage cheese which usually are consumed when freshly made. Also diacetyl is thought to be one of the four major basic compounds (with H_2S, dimethyl sulfide, and methanethiol) of Cheddar cheese flavor.[73] However, in view of the current complexities in what determines cheese flavor and how it is formed during ripening, this may not be too likely.

Maximum levels of diacetyl and indeed of all of these compounds occurs at the point where citrate utilization is just complete.[37,47,50,57] An obvious way to increase diacetyl production is to increase the citrate level which also results in prolongation of the time at which the peak level of diacetyl occurs.[57] The addition of citrate is often used in practice, e.g., in creaming mixtures for cottage cheese.[13-15]

In all types of mixed cultures, the level of diacetyl decreases once all the citrate is utilized.[40,47,49,50,57,74] This is due to increased synthesis and consequent increased activity of DR.[41] There are contrasting reports on acetoin reduction. Galesloot and Hassing[16] report no reduction in D cultures over a 14-day period and considerable reduction in B cultures while our results[57] suggest that reduction occurs in all types of cultures once citrate has been dissimilated; the reduction is more rapid in B cultures especially in the presence of Mn^{++}. Presumably, AR activity is responsible and 2,3-butylene glycol is formed, but this has not been positively determined. AR is present in both *S. lactis* subsp. *diacetylactis*[40] and *Leuconostoc* species,[41] though why it is apparently more active in leuconostocs is not obvious.

DR is present in some coliform and pseudomonads at much higher levels than in lactic acid bacteria.[74] These bacteria are potential contaminants of cultured buttermilk, cottage cheese, and sour cream and their growth in these products would inevitably contribute to a decrease in the diacetyl level and consequent lack of flavor.[74]

CO_2 production from citrate can be beneficial (eye formation in Gouda and Edam cheeses) or harmful (the cause of excessive openness in Cheddar cheese). The metabolism of lactose by *Leuconostoc* species also results in the production of 1 mol each of CO^2, lactate, and ethanol per mol of lactose metabolized. Production of 24 m*M* CO_2 from lactose (i.e., the theoretical maximum capable of being produced from citrate in milk) would also result in coproduction of 24 m*M* lactate (0.22%) or about one third of that produced by fully grown lactic starter cultures.[60] Little growth is needed before 0.22% lactate would be produced so that CO_2 production by the leuconostocs from lactose in milk may be as important as CO_2 production from citrate.

B. Acidity

Pure cultures of leuconostocs that produce diacetyl and acetoin will do so only at low pH in the presence of citrate or some other source of pyruvate.[52] Thus, in the preparation of good B cultures, one must ensure that the *Leuconostoc* species present in the mixed culture will not grow sufficiently rapidly to utilize all the citrate before the pH falls. pH is without any marked effect on diacetyl production by *S. lactis* subsp. *diacetylactis* since this organism produces diacetyl and acetoin as soon as it begins to metabolize citrate.[37,49,50] Leuconostocs are able to withstand low pH to a much better extent than the lactic streptococci.

C. Temperature

According to Cooper and Collins,[75] the minimum temperature for growth of *L. cremoris* is 14 to 17°C, the maximum is 34 to 36°C and the optimum is 24 to 27°C. The average generation time (GT) at 26°C is 3.2 hr. At 22°C the average generation time for the leuconostocs (3.5 hr) is more than double that of *S. cremoris* (1.5 hr). These data were obtained in complex broth, but larger differences in GT would probably exist in milk because of the inability of the leuconostocs to grow well in it. These workers concluded that such differences in GT could easily lead to domination of the leuconostocs by the streptococci in mixed cultures. Obviously other factors are involved because many mixed cultures are quite stable in commercial practice.

The prompt cooling of B and D cultures to 2°C prevents the reduction of the diacetyl level caused by DR activity.[47,76] The cooling must be carried out a point just before citrate utilization is complete since synthesis of DR and AR would still be repressed under such conditions; however, sufficient acid may not have been produced at this point in growth. Such a problem might be solved by a judicious increase in the level of citrate in the milk product. Pack et al.[76] found a greater number of flavor producers in a B and a D culture when incubated at 30°C than when incubated at 21°C, but greater amounts of diacetyl were produced at 21°C. This may be due to decreased activity of DR at 21°C or to slower growth of the flavor producers and, at least in the case of the B culture, the presence of more citrate to act as precursor at low pH.

D. Additional Nutrients

It has been accepted for a considerable time that a symbiosis exists between the lactic streptococci and the leuconostocs in mixed cultures, but the nature of the interaction has never been determined. It also is possible that most combinations of *S. lactis* or *S. cremoris* and leuconostocs would be incompatible suggesting the lack of involvement of any symbiosis, and that the B and BD cultures in use today were initially haphazard but stable mixtures that have evolved through continual use.

For optimum performance, cultures should be grown in a nutritious medium containing citrate and be free of inhibitors, especially bacteriophage and antibiotic residues (in the case of milk). For optimum growth in milk, leuconostocs need yeast extract (usually 0.3% w/v is sufficient); a few strains may also require glucose.[17] Manganese ions (1 to 5 $\mu g/m\ell$) will additionally stimulate leuconostocs in milk containing yeast extract but have no effect on *S. lactis* subsp. *diacetylactis*.[49,77] Mn^{++} also stimulates leuconostocs (and lactobacilli) in complex media.[37] Much lower concentrations of Mn^{++} (0.025 to 0.05 $\mu g/m\ell$) are required to stimulate *Leuconostoc* species in mixed cultures, but the leuconostocs die out more rapidly in the presence of the added Mn^{++} suggesting that it should be added judiciously.[77] Added Mn^{++} also increases the ability of B cultures to produce diacetyl and acetoin, presumably because of its stimulatory effect on growth rather than on the enzymes involved.[57,77] This effect of Mn^{++} is not without some practical significance since Menger et al.[78] have shown that the amount of citrate utilization by B cultures is exactly equal to the Mn^{++} content of milk. Both parameters were lowest in spring when the feed is lower in Mn^{++} and highest in autumn. Because of this, it is felt that the lower diacetyl levels in lactic butter made with B cultures in the spring is a direct result of a low Mn^{++} level in the milk and consequent inability of the leuconostocs to grow adequately.[79]

E. Green Flavor Defect

Many lactic cultures develop a "green" or yogurt flavor due to overproduction of acetaldehyde compared to diacetyl by either *S. lactis* subsp. *diacetylactis* or high acetaldehyde producing lactic streptococci combined with numbers of leuconostocs.[80] A green flavor develops if the ratio of diacetyl to acetaldehyde is $\leqslant$ 3:1. *Leuconostocs* have the ability to reduce acetaldehyde to ethanol and thus can be used to combat the green flavor defect.[81,82] Further evidence for this role of the leuconostocs is obtained on comparing acetaldehyde production by different mixed cultures.[57] All cultures exhibit production-destruction patterns for acetaldehyde, but the peak levels of acetaldehyde were lowest in B cultures (~1.0 μg/mℓ) followed in order by BD (~4.0 μg/mℓ) and D cultures (5.0 μg/mℓ). All cultures also exhibit production-destruction patterns for diacetyl, but the times of peak production for both acetaldehyde and diacetyl are different.[57,83] Thus the ratio of diacetyl to acetaldehyde will vary during growth suggesting that the optimum ratio of 4:1 found by Lindsay et al.[80] for a good flavored culture is not an absolute figure and should be used only as a guideline.

Acetaldehyde (100 μg/mℓ) has been reported[84] to stimulate the growth of *L. cremoris* and cause a twofold increase in the production of diacetyl and acetoin. Radiochemical data showed that neither diacetyl or acetoin were labeled. Whether a 100-fold smaller concentration of acetaldehyde (the approximate amount produced in B cultures) would have the same effect was not determined. However, it is not likely that it would.

F. Inhibitor Production

Several groups of workers have reported the inhibition of spoilage and/or pathogenic bacteria either by culture filtrates or when grown in assocation with mixed strain starter cultures or their individual component strains. These aspects have been reviewed by Speck[85] and Babel.[86] The determination of the exact nature of many of the produced inhibitors has proved elusive, however, certain factors have been identified as being involved including: (1) organic acids (notably acetic) which are especially inhibitory at low pH, (2) production of H_2O_2, (3) production of D-amino acids (like D-leucine), (4) depletion of essential nutrients, and (5) production of antibiotics like nisin.

G. Bacteriophage

The most important problem associated with the production of fermented dairy products is inhibition of acid production by the culture because of bacteriophage attack. Considerable attention has been given to the isolation, identification, and control of bacteriophage that attack the acid producers, however, very little attention has been given to those attacking the flavor producers, especially the leuconostocs. In fact, there are only two reports of the isolation of bacteriophage for the leuconostocs in dairy products.[87,88] Starter, whey, buttermilk, and cheese were the sources from which the bacteriophage were isolated but presumably the real source(s) were the leuconostocs themselves, i.e., lactic streptococci. No data were given of their effect on cultures; presumably they would destroy the CO_2 and diacetyl-producing abilities of cultures, as occurs in *S. lactis* subsp. *diacetylactis*.[89]

REFERENCES

1. Boekhout, F. W. J. and Ott de Vries, J. J., Aramabildner bei der Rahmsauerung, *Zentralbl. Bakteriol.*, 49, 373, 1919.
2. Storch, V., Fortsatte Undersfgelsen over Fremstillingen at Syravaekkere, 102 de Beretning fra Forsogslaboratoriet, Copenhagen, Denmark.
3. Hammer, B. N. and Bailey, D. E., The volatile acid production of starters and of organisms isolated from them, *Iowa State Coll. Res. Bull.*, 55, 1919.
4. Matuszewski, T., Pijanowski, E., and Supinska, J., *Streptococcus diacetilactis* u. sp. i. jego zastosowanie przy wryobie masla, *Pol. Agric. Forest Annu.*, 36, 1, 1936.
5. van Beynum, J. and Pette, J. W., De beteekenis der melkzuur bacterien bij de boteraromavorming, *Versl. Landbouwkd. Onderz.*, 42, 361, 1936.
6. Swartling, P. F., Biochemical and serological properties of some citric acid fermenting streptococci from milk and dairy products, *J. Dairy Res.*, 18, 256, 1951.
7. Deibel, R. H. and Seeley, H. W., Jr., The genus Streptococcus, in *Bergey's Manual of Determinative Bacteriology*, 8th ed., Buchanan, R. E. and Gibbons, N. E., Eds., Williams & Wilkins, Baltimore, 1974, 504.
8. Sherwood, I. R., The relation of certain lactic acid bacteria to open texture in Cheddar cheese, *J. Dairy Res.*, 10, 326, 1939.
9. Overcast, W. W. and Albrecht, T. W., Gas production in Cheddar cheese caused by *Leuconostoc citrovorum*, *J. Dairy Sci.*, 35, 554, 1952.
10. Galesloot, Th. E., An early gas defect in cheese caused by *Betacoccus* (*Leuconostoc*) species, *Neth. Milk Dairy J.*, 4, 274, 1950.
11. Sandine, W. E., Elliker, P. R., and Anderson, A. W., Taxonomic study of high carbon dioxide producing lactic acid streptococci isolated from mixed-strain starter cultures, *J. Dairy Sci.*, 42, 799, 1959.
12. Cogan, T. M., The utilization of citrate by lactic acid bacteria in milk and cheese, *Dairy Ind.*, 41, 12, 1976.
13. Mather, D. W. and Babel, F. J., A method for standardizing the biacetyl content of creamed cottage cheese, *J. Dairy Sci.*, 42, 1045, 1959.
14. Lunstedt, E. and Fogg, W. B., Citrated whey starters. II. Gradual formation of flavor and aroma in creamed cottage cheese after addition of small quantities of citrated cottage cheese whey cultures of *Streptococcus diacetylactis*, *J. Dairy Sci.*, 45, 1327, 1962.
15. Gilliland, S. E., Anna, E. D., and Speck, M. L., Concentrated cultures of *Leuconostoc citrovorum*, *Appl. Microbiol.*, 19, 890, 1970.
16. Galesloot, Th. E. and Hassing, F., Eukele verschillen in gedrag tussen zuursels met als aromabacterie *Streptococcus diacetilactis* of *Betacoccus cremoris*, *Neth. Milk Dairy J.*, 15, 212, 1961.
17. Garvie, E. I., The genus *Leuconostoc* and its nomenclature, *J. Dairy Res.*, 27, 283, 1960.
18. Garvie, E. I., The growth factor and amino acid requirements of species of the genus *Leuconostoc*, including *Leuconostoc paramesenteroides* (sp. nov) and *Leuconostoc oenos*, *J. Gen. Microbiol.*, 48, 439, 1967.
19. Garvie, E. I., *Leuconostoc oenos* sp. nov., *J. Gen. Microbiol.*, 48, 431, 1967.
20. Garvie, E. I., Hybridization between the deoxyribonucleic acids of some strains of heterofermentative lactic acid bacteria, *Int. J. Syst. Bacteriol.*, 26, 116, 1976.
21. Hontebeyrie, M. and Gasser, F., Deoxyribonucleic acid homologies in the genus *Leuconostoc*, *Int. J. Syst. Bacteriol.*, 27, 9, 1977.
22. Schlegel, J. A. and Babel, F. J., The citric acid fermenting bacteria associated with lactic cultures, in *Developments in Industrial Microbiology*, Vol. 7., Kuda, C. F., Ed., American Institute of Biological Sciences, Washington, D.C., 1966, chap. 30.
23. Stadhouders, J., Dairy starter cultures, *Milchwissenschaft*, 29, 329, 1974.
24. Abd-el-Malek, Y. and Gibson, T., Studies on the bacteria of milk. I. The streptococci of milk, *J. Dairy Res.*, 15, 233, 1948.
25. Marier, J. R. and Boulet, M., Direct determination of citric acid in milk with an improved pyridine-acetic anhydride method, *J. Dairy Sci.*, 41, 1683, 1958.
26. Westerfeld, W. W., A colorimetric determination of blood acetoin, *J. Biol. Chem.*, 161, 495, 1945.
27. Lawrence, R. C., Thomas, T. D., and Terzaghi, B. E., Reviews of the progress of dairy science: cheese starters, *J. Dairy Res.*, 43, 141, 1976.
28. Keenan, T. W., Production of acetic acid and other volatile compounds by *Leuconostoc citrovorum* and *Leuconostoc dextranicum*, *Appl. Microbiol.*, 16, 1881, 1968.
29. Goel, M. C. and Marth, E. H., Growth of *Leuconostoc citrovorum* in skim-milk at 22 and 30°C, *J. Dairy Sci.*, 52, 1207, 1969.

30. Waes, G., The enumeration of aromabacteria in BD starters, *Neth. Milk Dairy J.*, 22, 29, 1968.
31. Galesloot, Th. E., Hassing, F., and Stadhouders, J., Agar media voor het isoleren entellen van aromabacterien in zuursels, *Neth. Milk Dairy J.*, 15, 127, 1961.
32. Nickels, C. and Leesment, H., Methode zur Differenzierung und quantitativen Bestimmung von Saureweckerbakterien, *Milchwissenschaft*, 19, 374, 1964.
33. deMan, J. C., Rogosa, M., and Sharpe, M. E., A medium for the cultivation of lactobacilli, *J. Appl. Bacteriol.*, 23, 130, 1960.
34. Reddy, M. S., Vedamuthu, E. R., Washam, C. J., and Reinbold, G. W., Agar medium for differential enumeration of lactic streptococci, *Appl. Microbiol.*, 24, 947, 1972.
35. Kempler, G. M. and McKay, L. L., Improved medium for detection of citrate-fermenting *Streptococcus lactis* subsp. *diacetylactis*, *Appl. Environ. Microbiol.*, 39, 926, 1980.
36. Fryer, T. F., Utilization of citrate by lactobacilli isolated from dairy products, *J. Dairy Res.*, 37, 9, 1970.
37. Drinan, D. F., Tobin, S., and Cogan, T. M., Citric acid metabolism in hetero- and homofermentative lactic acid bacteria, *Appl. Environ. Microbiol.*, 31, 481, 1976.
38. Harvey, R. J. and Collins, E. B., Citrate transport system of *Streptococcus diacetilactis*, *J. Bacteriol.*, 83, 1005, 1962.
39. Harvey, R. J. and Collins, E. B., The citratase of *Streptococcus diacetilactis*. Substrate, products, equilibrium, *J. Biol. Chem.*, 238, 2648, 1963.
40. Cogan, T. M., Constitutive nature of the enzymes of citrate metabolism in *Streptococcus lactis* subsp. *diacetylactis*, *J. Dairy Res.*, 48, 489, 1981.
41. Mellerick D. and Cogan, T. M., Induction of some enzymes of citrate metabolism in *Leuconostoc lactis* and other hetero-fermentative lactic acid bacteria, *J. Dairy Res.*, 48, 497, 1981.
42. Speckman, R. A. and Collins, E. B., Diacetyl biosynthesis in *Streptococcus diacetilactis* and *Leuconostoc citrovorum*, *J. Bacteriol.*, 95, 174, 1968.
43. Seitz, E. W., Sandine, W. E., Elliker, P. R., and Day, E. A., Studies on diacetyl biosynthesis by *Streptococcus diacetilactis*, *Can. J. Microbiol.*, 8, 161, 1962.
44. Collins, E. B. and Bruhn, J. C., Roles of acetate and pyruvate in the metabolism of *Streptococcus diacetilactis*, *J. Bacteriol.*, 103, 541, 1970.
45. Bryn, E., Hetland, O., and Stormer, F. C., The reduction of diacetyl and acetoin in *Aerobacter aerogenes*. Evidence for one enzyme catalysing both reactions, *Eur. J. Biochem.*, 18, 116, 1971.
46. Collins, E. B. and Speckman, R. A., Evidence for cellular control in the synthesis of acetoin or α-ketoisovaleric acid by microorganisms, *Can. J. Microbiol.*, 20, 805, 1976.
47. Jönsson, H. and Pettersson, H. E., Studies on the citric acid fermentation in lactic starter cultures with special interest in α-acetolactic acid. II. Development of analytical procedures and metabolic studies, *Milchwissenschaft*, 32, 513, 1977.
48. Jönsson, H. and Pettersson, H. E., Studies on the citric acid fermentation in lactic starter cultures with special interest in α-acetolactic acid. I. Materials and methods, *Milchwissenschaft*, 32, 513, 1977.
49. Cogan, T. M., Citrate utilization in milk by *Leuconostoc cremoris* and *Streptococcus diacetilactis*, *J. Dairy Res.*, 42, 139, 1975.
50. Cogan, T. M., Acetoin production and citrate metabolism in *Streptococcus lactis* subsp. *diacetylactis*, *Ir. J. Food Sci. Technol.*, 6, 69, 1982.
51. Harvey, R. J. and Collins, E. B., Roles of citrate and acetoin in the metabolism of *Streptococcus diacetilactis*, *J. Bacteriol.*, 86, 1301, 1963.
52. Cogan, T. M., O'Dowd, M., and Mellerick D., Effects of pH and sugar on acetoin production from citrate by *Leuconostoc lactis*, *Appl. Environ. Microbiol.*, 41, 1, 1981.
53. van Beynum, J. and Pette, J. W., The decomposition of citric acid by *Betacoccus cremoris*, *J. Dairy Res.*, 10, 250, 1939.
54. Michaelian, M. B., Hoecker, W. H., and Hammer, B. W., Effect of pH on the production of acetylmethylcarbiol plus diacetyl in milk by the citric acid fermenting streptococci, *J. Dairy Sci.*, 21, 213, 1938.
55. Cox, G. A., The effect of acidity on the production of diacetyl by betacocci in milk, *J. Dairy Res.*, 14, 28, 1945.
56. deMan, J. C., Over de wijze waarop diacetyl in culturen van *Betacoccus cremoris* ontstaat, *Neth. Milk Dairy J.*, 10, 38, 1956.
57. Walsh, B. and Cogan, T. M., Diacetyl, acetoin and acetaldehyde production by mixed species lactic starter cultures, *Appl. Microbiol.*, 26, 82, 1973.
58. Galesloot, Th. E., The bacteriology and biochemistry of starters and ripened cream, *16th Int. Dairy Congr.: D*, 143, 1962.
59. Collins, E. B., Biosynthesis of flavor compounds by microorganisms, *J. Dairy Sci.*, 55, 1022, 1972.
60. Cogan, T. M., Determination of growth rates of lactic starter cultures, *Ir. J. Food Sci. Technol.*, 2, 105, 1978.

61. Dills, S. S., Aperson, A., Schmidt, M. R., and Saier, M. H., Jr., Carbohydrate transport in bacteria, *Microbiol. Rev.*, 44, 385, 1980.
62. Romano, A. H., Trifone, H. D., and Brustolon, M., Distribution of the phosphoenol pyruvate: glucose phosphotransferase system in fermentative bacteria, *J. Bacteriol.*, 139, 93, 1979.
63. Thomas, T. D., Regulation of lactose fermentation in Group N streptococci, *Appl. Environ. Microbiol.*, 32, 474, 1976.
64. Thompson, J. and Thomas, T. D., Phosphoenol pyruvate and 2-phosphoglycerate: endogenous energy source(s) for sugar accumulation by starved cells of *Streptococcus lactis*, *J. Bacteriol.*, 130, 583, 1977.
65. Thompson, J. Turner, K. W., and Thomas, T. D., Catabolite inhibition and sequential metabolism of sugars by *Streptococcus lactis*, *J. Bacteriol.*, 113, 1163, 1978.
66. Thompson, J., In vivo regulation of glycolysis and characterization of sugar: phosphotransferase systems in *Streptococcus lactis*, *J. Bacteriol.*, 136, 465, 1978.
67. Thomas, T. D., Ellwood, D. C., and Longyear, V. M. C., Change from homo- to heterolactic fermentation by *Streptococcus lactis* resulting from glucose limitation in anaerobic chemostat cultures, *J. Bacteriol.*, 138, 109, 1979.
68. Thompson, J. and Chassy, B. M., Uptake and metabolism of sucrose by *Streptococcus lactis*, *J. Bacteriol.*, 147, 543, 1981.
69. Davies, F. L. and Gasson, M. J., Reviews of the progress of dairy science: genetics of lactic acid bacteria, *J. Dairy Res.*, 48, 363, 1981.
70. Kempler, G. M. and McKay, L. L., Biochemistry and genetics of citrate utilization in *Streptococcus lactis* subsp. *diacetilactis*, *J. Dairy Sci.*, 64, 1527, 1981.
71. Kempler, G. M. and McKay, L. L., Characterization of plasmid deoxyribonucleic acid in *Streptococcus lactis* subsp. *diacetylactis:* evidence for plasmid-linked citrate utilization, *Appl. Environ. Microbiol.*, 37, 316, 1979.
72. Hempenius, W. H., Liska, B. J., and Harrington, R. B., Consumer research on flavor levels in sour cream, *J. Dairy Sci.*, 50(Abstr.), 954, 1967.
73. Manning, D. J. and Robinson, H. M., The analysis of volatile substances associated with Cheddar cheese aroma, *J. Dairy Res.*, 40, 63, 1973.
74. Seitz, E. W., Sandine, W. E., Elliker, P. R., and Day, E. A., Distribution of diacetyl reductase among bacteria, *J. Dairy Sci.*, 46, 186, 1963.
75. Cooper, R. K. and Collins, E. B., Influence of temperature on growth of *Leuconostoc cremoris*, *J. Dairy Sci.*, 61, 1085, 1978.
76. Pack, M. Y., Vedamuthu, E. R., Sandine, W. E., Elliker, P. R., and Leesment, H., Effect of temperature on growth and diacetyl production by aroma bacteria in single- and mixed-strain lactic cultures, *J. Dairy Sci.*, 51, 339, 1968.
77. deMan, J. C. and Galesloot, Th. E., De involed van een toevoeging van mangaan de milk op de groei van zuurselsbacteriën, *Neth. Milk Dairy J.*, 16, 1, 1962.
78. Menger, J. W., Das, H. A., and van Raaphorst, J. G., The manganese content of milk. A discussion on its relation to the bacteriological composition of starters containing *Leuconostoc* (Betacoccus) *cremoris*, *Neth. Milk Dairy J.*, 21, 67, 1967.
79. Galesloot, Th. E. and Hassing, F., Verdere onderzoekingen betreffende de involed van het jaargetijde op het gedrag van zuursels, *Neth. Milk Dairy J.*, 16, 24, 1962.
80. Lindsay, R. C., Day, E. A., and Sandine, W. E., Green flavor defect in lactic starter cultures, *J. Dairy Sci.*, 48, 863, 1965.
81. Keenan, T. W., Lindsay, R. C., and Day, E. A., Acetaldehyde utilization by Leuconostocs, *Appl. Microbiol.*, 14, 802, 1966.
82. Keenan, T. W. and Lindsay, R. C., Removal of green flavor from ripened butter cultures, *J. Dairy Sci.*, 49, 1563, 1966.
83. Keenan, T. W., Lindsay, R. C., Morgan, M. E., and Day, E. A., Acetaldehyde production by single strain lactic streptococci, *J. Dairy Sci.*, 49, 10, 1966.
84. Collins, E. B. and Speckman, R. A., Influence of acetaldehyde on growth and acetoin production by *Leuconostoc citrovorum*, *J. Dairy Sci.*, 57, 1428, 1974.
85. Speck, M. L., Control of food-borne pathogens by starter cultures, *J. Dairy Sci.*, 55, 1019, 1972.
86. Babel, F. J., Antibiosis by lactic culture bacteria, *J. Dairy Sci.*, 60, 815, 1977.
87. Sozzi, T., Paulin, J. M., Maret, R., and Pousaz, R., Isolation of a bacteriophage of *Leuconstoc mesenteriodes* from dairy products, *J. Appl. Bacteriol.*, 44, 159, 1978.
88. Shin, C. and Sato, Y., Isolation of leuconostoc bacteriophages from dairy products, *Jpn. J. Zootech. Sci.*, 50, 419, 1979.
89. Sandine, W. E., Elliker, P. R., and Hays, H. A., Bacteriophage-lysis of *Streptococcus diacetylactis* and its effect on diacetyl production in mixed-strain starter cultures, *J. Dairy Sci.*, 43, 755, 1960.
90. Fitzgerald, R. J., Doonan, S., and Cogan, T. M., Unpublished data.
91. O'Sullivan, T., M.S. thesis, University College, Cork, Ireland, 1982.

Chapter 4

THE LACTOBACILLI: MILK PRODUCTS

Stanley E. Gilliland

TABLE OF CONTENTS

I. SPECIES INVOLVED

The primary species of lactobacilli used as starter cultures in the manufacture of cultured milk products are homofermentative. *Lactobacillus bulgaricus* and *L. lactis* are the species most often encountered. Products in which these are used include yogurt, Swiss cheese, Italian cheese, Emmental cheese, Bulgarian buttermilk, kefir, and koumiss.[1] *L. helveticus* is also involved, in some cases, in some of these products.[1] *L. acidophilus* is used in the manufacture of acidophilus milk, nonfermented acidophilus milk, and acidophilus yogurt.[2,3] *L. casei* is involved as a starter culture in the production of a Japanese product called Yakult.[2]

Bergey's Manual of Determinative Bacteriology[4] contains a thorough description of the identity characteristics of the lactobacilli. All of the homofermentative lactobacilli are Gram-positive, catalase-negative, nonspore forming, rod-shaped bacteria. The major identification characteristics of these bacteria important in dairy starter cultures are given in Table 1. Differentiation among species of lactobacilli relies heavily on their ability or lack of ability to ferment a number of carbohydrates. This is quite often a cumbersome technique; however, new miniaturized systems that have been developed for microbial identification have proven to be very useful in this regard.[5] Gilliland and Speck[5] were successful in differentiating among several species of lactobacilli using the Minitek system (Baltimore Biological Laboratory, Cockeysville, Md.). This system involves the use of small filter paper discs impregnated with the dried substrate and indicator. The sterile discs containing the desired substrates are aseptically placed in specially designed plastic plates having individual wells for each disc. The lactobacilli collected from colony(s) on agar surface and dispensed in a basal broth (devoid of sugars) are added to the wells containing the discs. After anaerobic incubation the "fermentation patterns" are recorded and compared to those presented in *Bergey's Manual of Determinative Bacteriology*[4] for identification. The "fermentation patterns" combined with the ability of the lactobacilli to grow at 15°C provides the basis for their identification. The substrates used included amygdalin, arabinose, arginine, cellobiose, esculin, galactose, glucose, lactose, maltose, mannitol, mannose, melezitose, melibiose, raffinose, rhamnose, salicin, sorbitol, sucrose, trehalose, and xylose.

Classification groups described by Rogosa and Sharpe[6] are often used in describing the lactobacilli. These classification groups include the Thermobacterium, Streptobacterium, and Betabacterium. The first two of these groups include the homofermentative lactobacilli and the last group is composed entirely of the heterofermentative species. The differentiation between the Thermobacterium and the Streptobacterium groups is based largely on the differences in growth temperatures of the species involved. In general, members of the Thermobacterium group exhibit higher optimum growth temperatures than those in the Streptobacterium group. Of the homofermentative lactobacilli involved as starters in dairy products, *L. casei* is a member of the Streptobacterium group and the others are members of the Thermobacterium group. The ability of the lactobacilli to grow or not grow at 15°C appears to be more important for identification of the lactobacilli than does the ability of the organism to grow at 45°C.

The colonial appearance of lactobacilli on the surface of agar medium is of limited value as an identity characteristic.[7] This is especially true since it has been shown that a single culture may yield a mixture of types of colonies. Wheater[7] reported that a pure culture isolated from a single colony, when restreaked on agar medium, often yielded more than one colony type. Thus, care must be exercised in differentiating the lactobacilli based on colonial morphology.

While the lactobacilli must possess at least some proteolytic activity in order to grow

Table 1
CHARACTERISTICS IMPORTANT IN DIFFERENTIATING LACTOBACILLI NORMALLY INVOLVED AS STARTER BACTERIA IN MILK PRODUCTS[a]

	L. lactis	*L. bulgaricus*	*L. helveticus*	*L. acidophilus*	*L. casei-casei*	*L. casei-rhamnosus*
Fermentation						
Amygdalin	−	−	−	+	+	+
Arabinose	−	−	−	−	−	−
Cellobiose	−	−	−	+	+	+
Galactose	+	+	+	+	+	+
Glucose[b]	+	+	+	+	+	+
Lactose	+	+	+	+	+	+
Maltose	+	−	+	+	±	±
Mannitol	−	−	−	−	+	+
Mannose	+	−	W	+	+	+
Melezitose	−	−	−	−	+	+
Melibiose	−	−	−	±	−	−
Raffinose	−	−	−	±	−	−
Rhamnose	−	−	−	−	−	+
Salicin	+	−	−	+	+	+
Sorbitol	−	−	−	−	+	+
Sucrose	+	−	−	+	±	±
Trehalose	+	−	±	+	+	+
Xylose	−	−	−	−	−	−
Hydrolysis of esculin	W[c]	−	−	+	+	+
NH_3 from arginine	−	−	−	−	−	−
Growth at 15°C	−	−	−	−	+	+
Growth at 45°C	+	+	+	+	−	+

[a] According to *Bergey's Manual of Determinative Bacteriology*.[4]
[b] None produce gas.
[c] W: Implies weak or negative reation.

in a product such as milk, they are not sufficiently proteolytic to be so classified with regard to identity characteristics. They also are considered to be nonlipolytic.

Considerable effort has been directed toward designating serological groups of the lactobacilli. In 1955, Sharpe[8] proposed six serological groups. There was, however, some overlap of species within the serological groups in that some species having major biochemical differences, based on fermentation patterns, were included in common groups. The antigenic composition varies within species and has not been completely characterized for all species.[9] Cells of given species of lactobacilli may contain more than one antigen.[10] The antigenic components that have been identified include primarily polysaccharide-type compounds containing components such as α-glucopyranose, glucose, galactose, rhamnose, n-acetylglucosamine, and phosphate.[9-11] While the antigenic make-up of the lactobacilli has not been completely determined, in the future it may provide a very useful tool for classifying the lactobacilli.

II. FUNCTIONS AS STARTER CULTURES

The overall effect of the starter culture used in the manufacture of many dairy products is due to the presence and action of more than one species of bacteria. The lactobacilli are often used in mixed species starter cultures along with other species such as *Streptococcus thermophilus* and *Propionibacterium shermanii.* The major contribution of the lactobacilli is the production of lactic acid that results from the fermentation of lactose. Even though the species involved are primarily homofermentative they do produce significant amounts of metabolites other than lactic acid. Volatile compounds produced by the lactobacilli during growth in milk include acetaldehyde, diacetyl, and alcohol.[12-14] The major volatile compound produced is acetyaldehyde. In addition, *L. casei* produces significant amounts of diacetyl from the fermentation of citrate. The supplementation of milk with citrate enhances diacetyl production by this organism. While diacetyl produced by this species of lactobacilli may contribute to the flavor of cultured products, *L. casei* is not normally the culture involved in products where the major flavor component is diacetyl.

Acetaldehyde has been identified as a major flavor component of yogurt. A desirable typical flavor of this product requires 8 ppm acetaldehyde.[14] The lactobacilli in yogurt starter cultures have been identified as the main source of this acetaldehyde. They also produce detectable amounts of acetone that can influence the flavor of yogurt. Lactobacilli used for yogurt have been divided into three groups with regard to the relative amounts of acetaldehyde and acetone produced (Table 2). An acetaldehyde to acetone ratio of $\geq$2.8:1 is recommended for the best yogurt flavor.[14] Some acetaldehyde may be produced from sugar metabolism; however, in *L. bulgaricus* a considerable amount of acetaldehyde is produced from the degradation of threonine via the action of threonine aldolase.[15] Certain of the lactobacilli are also capable of reducing acetaldehyde to form ethanol, however, the amounts of ethanol produced are quite variable.[12] While the ethanol may contribute to the overall flavor of the product it should be considered a component of minor importance.

Taxonomically, the lactobacilli are not considered to be proteolytic, although in order to grow in milk they must be sufficiently proteolytic in order to obtain the needed nitrogen from the milk proteins. The proteolytic activity of these microorganisms may well contribute to the texture and flavor properties of cultured milk products in which they are used as starter cultures. The proteolytic activity may also result in the release of sufficient free amino acids to stimulate the growth and action of other organisms used in mixed species starter cultures in conjunction with the lactobacilli.[16]

There is proteolytic activity associated with the cell wall and peptidase activity in the

Table 2
ACETALDEHYDE, ACETONE, ACETOIN, AND DIACETYL PRODUCTION BY THERMOPHILIC LACTOBACTILLI AND TYPE OF FLAVOR

Groups	Strains examined (%)	Average amounts of Acetaldehyde (ppm)	Acetone (ppm)	Acetoin	Diacetyl	Flavor type
A	22.6	1.43	3.24	Trace	—	Slight and not typical
B	25.2	3.88	2.44	Trace	—	Clean, but not very full
C	52.2	8.50	3.00	Trace	—	Very good and strong

From Bottazzi, V. and Vescovo, M., *Neth. Milk Dairy J.*, 23, 71, 1969. With permission.

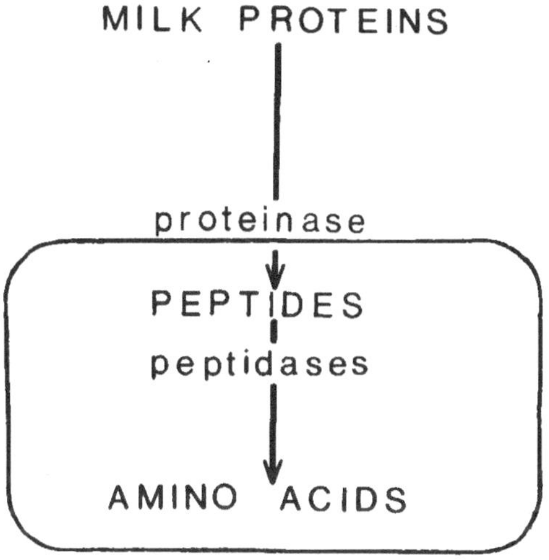

FIGURE 1. Function of proteinase(s) and intracellular peptidase(s) in the utilization of milk proteins by the lactobacilli.

intracellular portions of the lactobacilli.[17,18] The surface-bound enzymes are responsible for hydrolyzing or partially hydrolyzing large protein molecules sufficiently to yield peptides small enough to be transported into the cell where the peptidases further degrade them to the individual amino acids needed for growth of the organism (Figure 1). The proteolytic activity of the lactobacilli is optimum at temperatures of 45 to 50°C and at pH 5.2 to 5.8.[17] It is greatly reduced at temperatures greater than 55°C. This has been attributed to the thermodenaturation of the enzyme.[17] The enzyme was completely inactivated by heating for 1 min at 70°C. The proteinase of *L. bulgaricus* growing in milk is most active on β-casein followed by α_s and κ casein.[19] Very little activity was demonstrated on whey proteins. Both the surface-bound proteinase and the intracellular peptidase enzymes are very important in the area of cheese ripening.[20,21] The intracellular peptidase or protease of *L. bulgaricus* and *L. helveticus* is more active on α_s casein than on either κ or β-caseins.[22] Thus, there appears to be some difference in the specificity of the surface-bound proteinase and the intracellular protease of these organisms with regard to the types of casein they preferentially attack. The intracellular proteases would become important during cheese ripening as the cells of the lactobacilli lyse during the ripening process.

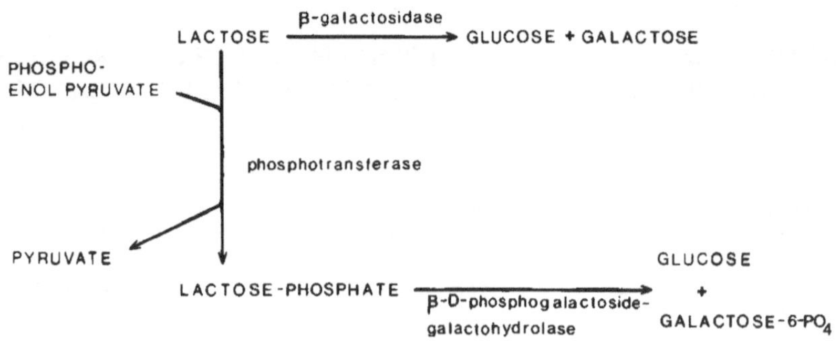

FIGURE 2. Lactose hydrolyzing reactions used by lactobacilli.

In order to obtain energy for growth in milk, the homofermentative lactobacilli must be able to utilize lactose. Even though the fermentation of lactose by the lactobacilli is their most important function in the manufacture of cultured milk products, the lactose hydrolyzing system(s) of these organisms have not received as much attention in research as have such systems in the streptococci. The lactobacilli hydrolyze lactose by the action of β-galactosidase and/or β-D-phosphogalactoside galactohydrolase (Figure 2). Most homofermentative lactobacilli possess the enzyme β-galactosidase.[23-25] However, in *L. casei* the lactose hydrolyzing enzyme is β-D-phosphogalactoside galactohydrolase.[24,25] This enzyme requires that lactose be phosphorylated prior to being transported into the cell. Hydrolysis of lactose occurs within the cells for both of the enzymes. *L. lactis* and *L. helveticus* contain both enzymes, however, they both possess greater β-galactosidase activity than β-D-phosphogalactoside galactohydrolase activity.[25] At present β-galactosidase appears to be the most important of the two enzymes. In *L. bulgaricus* it has optimum activity at pH 7 and is activated by cations such as magnesium, manganese, and iron.[23,26]

Given the option, the lactobacilli would in most cases use glucose as an energy source rather than lactose or galactose.[27] In some lactobacilli, free galactose is liberated from cells growing on lactose.[24] This results in galactose accumulating in milk cultures for such organisms.

After the initial hydrolysis of lactose in the cells of the homofermentative lactobacilli the hexose components are fermented via the Embden Meyerhof pathway[28-30] (Figure 3). The heterofermentative lactobacilli on the other hand ferment the hexoses via the hexose monophosphate pathway. The heterofermentative lactobacilli lack the enzymes aldolase and triosephosphate isomerase which are key enzymes in the Embden Meyerhof pathway.[28,29] The presence of these enzymes in homofermentative lactobacilli and their absence in the heterofermentative species constitutes one of the major differences between these two groups of bacteria. Many of the other enzymes involved in the Embden Meyerhof pathway are also found in heterofermentative lactobacilli.

For the galactose moiety of lactose to be fermented requires that the sugar be converted to a phosphorylated derivative of glucose.[28,31] Galactose is converted through several steps to glucose-6-phosphate which is further metabolized via the Embden Meyerhof pathway (Figure 4).

The net result of the Embden Meyerhof pathway is to yield energy in the form of two molecules of adenosine triphosphate (ATP) for each molecule of glucose. This supplies the energy needed for growth of the lactobacilli. The fermentation of galactose yields only one ATP per molecule because of the energy (1 ATP) required to convert galactose to glucose-6-phosphate. The formation of less energy from galactose than from glucose may explain why some lactobacilli accumulate free galactose during growth in milk.

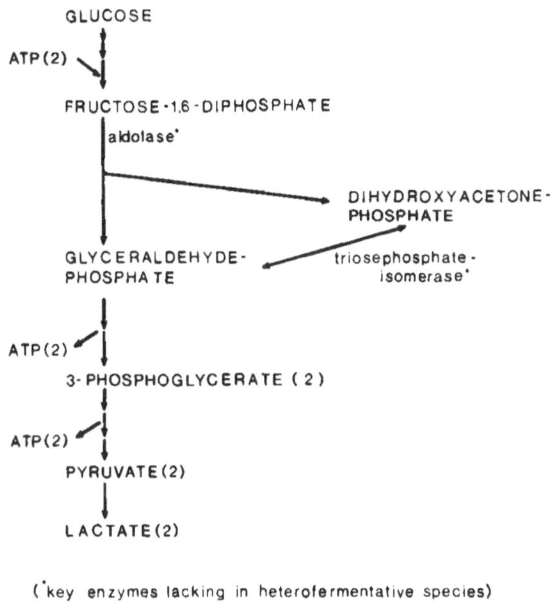

FIGURE 3. Embden Meyerhof pathway used by homofermentative lactobacilli for sugar fermentation.

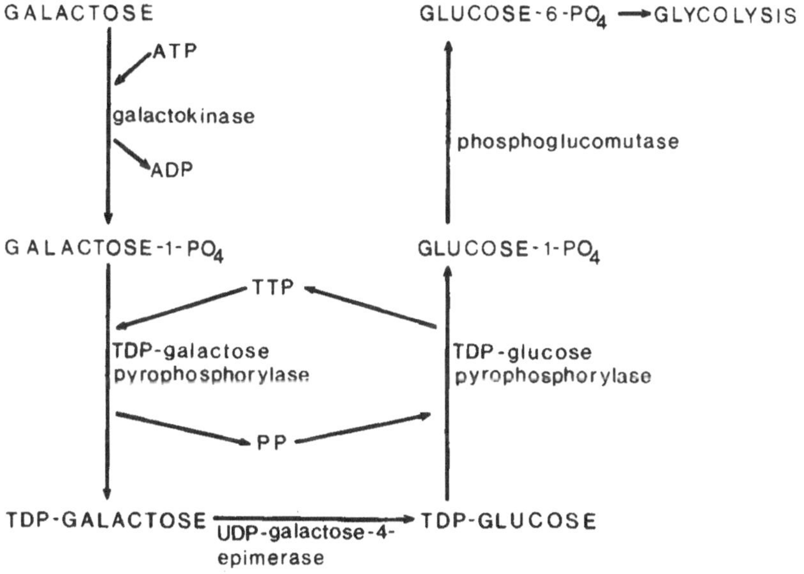

FIGURE 4. Pathway for conversion of galactose to glucose-6-phosphate.

The energy yielding fermentation pathways are important in the manufacture of cultured dairy products because lactic acid is the primary end product of each pathway. Thus, they are probably the most important metabolic pathways in the homofermentative lactobacilli in relation to their function in cultured milk products.

III. FACTORS INFLUENCING OPTIMUM PERFORMANCE AS STARTER CULTURES

As with other starter cultures, the proper performance of the lactobacilli requires that the culture receives proper care and handling between uses. The propagation procedures utilized in producing the starter culture are very important. Furthermore, variation among strains of the individual species of lactobacilli must be considered when evaluating optimum performance.

A. Temperature

The optimum performance of the lactobacilli is normally expected to coincide with the optimum growth temperature for each species. In some cases, however, it may be necessary to consider other factors. In the case of yogurt starter cultures, which are mixed species starters usually containing both *L. bulgaricus* and *S. thermophilus,* the two individual species of bacteria exhibit different optimum growth temperatures. Fortunately the temperature (41 to 42°C) recommended for maintaining the proper balance between the two species of the bacteria is close to that of the optimum growth temperature for *L. bulgaricus.*

B. Treatment of Milk Prior to Inoculation

Other than standardizing the milk with regard to total solids or fat content, the primary treatment that is utilized in preparing the milk for manufacture of cultured milk products is heat processing. This has several functions; among them is the killing of undesirable organisms that might have an adverse effect on the product being manufactured. It is generally accepted that heating of the milk for cultured products improves it as the growth medium for the starter culture bacteria. The effect of heat on the proteins of the milk apparently causes some alteration(s), making them easier to be utilized by the starter bacteria. Since the lactobacilli are microaerophilic in nature, heating of the milk may also play an important role in reducing the amount of oxygen present in the product. This is supported by the fact that the lactobacilli generally grow best in freshly heated milk. The heat process of milk may also destroy or at least partially destroy natural inhibitors occurring in raw milk.[32] If sufficiently high temperatures are utilized in heating the milk, some formic acid is formed. The presence of this compound in milk can result in stimulation of the growth of certain of the lactobacilli.[33]

C. Nutrient Availability

The general requirements for growth of the lactobacilli include water, fermentable sugar, carbon, nitrogen (including preformed amino acids), vitamins, and other cofactors. The organisms are considered to be fastidious in nature. All nutrients essential for growth of the lactobacilli are contained in milk. The problem is that the nutrients do not necessarily occur in the most easily used forms. For example, the fermentable sugar (lactose) is more difficult to utilize by some cultures than a monosaccharide such as glucose. The necessary amino acids occur in the form of proteins. Thus, the organism must be able to hydrolyze both lactose and at least partially hydrolyze milk protein in order to obtain the nutrients needed for growth. In some cases poor performance of the lactobacilli in producing the desired change(s) in the milk during the culturing process may be due to limitations in their ability to hydrolyze lactose and/or proteins.

L. bulgaricus and *S. thermophilus* used as yogurt cultures produce acid faster in milk in which the lactose had been prehydrolyzed by treatment with β-galactosidase.[34] In some cases however, stimulation is not obtained by prehydrolyzing the lactose.[35] This difference in response is probably due to variations in the cultures being tested. For those starters that can hydrolyze the lactose as rapidly as it can be used for growth and acid production prehydrolysis of lactose may not provide a benefit. In milk treated with β-galactosidase to prehydrolyze the lactose, *L. bulgaricus* uses only the glucose portion of the molecule.[27]

Some lactobacilli, when growing in a medium containing lactose as the sugar source, will excrete the galactose and not utilize it for growth.[24] These observations suggest that the lactobacilli, in addition to exhibiting a preference for glucose are also utilizing the hexose moiety of lactose that yields the most energy per molecule.

The nonprotein nitrogen fraction of milk would be inadequate to meet the needs of growing lactobacilli.[18] To grow exponentially in milk, the proteinase activity of the lactobacilli must function effectively in order for the organism to obtain sufficient amino acids for growth. Slow growing strains of lactobacilli may have inadequate proteinase activity. The growth of *L. casei* is slower when casein is the nitrogen source than when peptones are supplied to the culture.[20] During growth in milk, *L. bulgaricus* apparently exhibits a preference for utilizing β-casein as a nitrogen source over the other proteins, indicating that the type of protein is also an important factor in influencing the growth of these cultures.

D. Stimulatory Materials

In efforts to improve growth and acid production by lactic acid bacteria in milk, a number of substances have been evaluated for possible stimulatory effects over the years. Yeast extract, various peptones, liver extract, corn steep liquor, and tomato juice are some of the materials that have been found to contain substances that stimulate the growth and action of starter culture bacteria in milk. While such materials may be stimulatory when added to milk for cultured dairy products, their use in most cases has not been practical due to undesirable flavors that accompany them. In general, the stimulants in these products include such compounds as peptones, amino acids, vitamins, and other co-factors. Corn steep liquor, for example, contains amino acids and peptides that are stimulatory for *L. casei*.[36] Tomato juice contains both stimulants and inhibitors for the lactobacilli.[37,38] The stimulants in this material include both adenine and adenosine. The inhibitor is a nucleotide containing adenine and xylose. Pancreas extract has shown promise as a source of stimulants for starter cultures used in the manufacture of milk products since it does not impart an undesirable flavor in the milk. However, it is rather expensive and thus is not often utilized as a stimulant. The main stimulatory components from pancreas extract have been identified as peptides, especially when *L. casei* was used as the test organism.[39]

Other compounds stimulatory for the lactobacilli include oleic acid and formate.[40,41] Formate added to milk can reduce the stimulatory effect that *S. thermophilus* has toward *L. bulgaricus* during yogurt manufacture. Thornhill and Cogan[42] reported that certain fruit preparations added to yogurt resulted in increased acid production by the yogurt culture. They concluded, however, that such additions did not result in a major stimulation for the culture.

E. Metabolites Produced by Lactobacilli

As mentioned previously the major metabolite produced by the lactobacilli that makes them important in cultured milk products is lactic acid. Even though the lactobacilli are considered acid tolerant, sufficient lactic acid will limit their growth. It is possible that in some cases, variations observed among species and/or strains of lactobacilli with regard to the rapidity of growth and acid production in milk may be due to variations in their sensitivity to lactic acid.

The lactobacilli, being catalase negative, do not have the usual mechanism for destroying metabolically produced hydrogen peroxide. For this reason the cultures can accumulate hydrogen peroxide in the growth medium. Some lactic acid bacteria produced autoinhibitory levels of hydrogen peroxide.[43,44] The addition of catalase to milk cultures often stimulates acid production by lactic streptococci, but not by *L. acidophilus* and *L. bulgaricus*.[43] However, the lactobacilli produce more total peroxide than do the streptococci. This indicates that the lactobacilli are more tolerant to hydrogen peroxide than the streptococci. In another study on the evaluation of the possibility of adding cells of *L. acidophilus* to yogurt that had been cultured with normal yogurt cultures, the *L. acidophilus* cells died during storage due to hydrogen peroxide produced by *L. bulgaricus* in the yogurt culture.[45] It is possible that hydrogen peroxide produced by some strains of lactobacilli may limit their growth and acid production in milk. Most likely there is considerable variation among strains of different species with regard to this characteristic.

The inability of the lactobacilli to destroy metabolically produced hydrogen peroxide may, at least partially, explain their poor growth under highly aerobic conditions. This suggests the importance of avoiding excessive agitation when mixing the starter into the milk to be cultured. The inability to grow under highly aerobic conditions may also be related to oxygen toxicity. In many bacteria superoxide dismutase serves as a defense against oxygen toxicity, however, the lactobacilli do not contain this enzyme. Some lactobacilli contain dialyzable manganese that may replace the function of superoxide dismutase.[46] The intracellular manganese serves as an oxygen scavenger. This emphasizes the importance of providing adequate manganese for the culture during growth.

F. Interactions with Other Starter Bacteria

In yogurt starter cultures *L. bulgaricus* and *S. thermophilus* growing together produce acid more rapidly than either of the individual species alone.[16,47,48] This appears to be a two-way beneficial interaction. During growth in milk the *L. bulgaricus* cells produce sufficient free amino acids, particularly histadine, to stimulate growth and acid production by *S. thermophilus*.[16] The *S. thermophilus*, on the other hand, produces formic acid, which is stimulatory for *L. bulgaricus*.[49] Also, *S. thermophilus* produces sufficient amounts of carbon dioxide to stimulate the growth and action of *L. bulgaricus*.[50] In addition to the beneficial interaction between these two species, a competitive interaction has also been reported.[51] Even though acid production by both was greater than either one alone, the total growth of *L. bulgaricus* was less in the associative culture. This was apparently due to the streptococci utilizing some component essential for maximum growth of the lactobacilli.

G. Handling of Starter Culture Between Uses

As with other starter cultures the temperature at which the culture is stored influences its subsequent performance in milk.[47] Storage at −30 to 0°C causes reduced activity compared to storage at 14°C. Storage at 8°C results in a culture having intermediate activity. This suggests that refrigerated storage of the lactobacilli just prior to use may be slightly detrimental to their performance in the manufacture of cultured milk products. Smittle et al.[52] reported variation among strains with regard to survival of *L. bulgaricus* stored at −196°C. However, all strains of the lactobacilli grown in the presence of Tween® 80 or oleic acid survived freezing better than those grown without these materials. The improved survival was related to the level of a C_{19} cyclopropane fatty acid in the cellular lipids of the lactobacilli (Figure 5).[53] Growth of the organisms in the medium containing oleic acid resulted in increased levels of this fatty acid. Similar findings were reported with regard to storage of the lactobacilli at −17°C. Cells grown in a medium containing oleic acid survived freezing better than those grown without the fatty acid. The increased survival was attributed to an increased ratio of

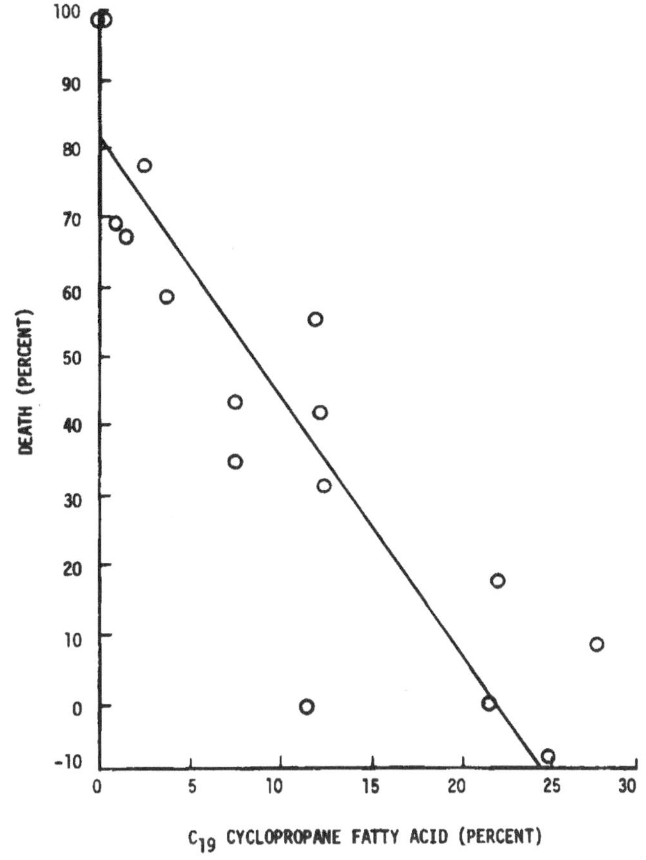

FIGURE 5. Regression of percent death (during freezing at −196°C) on percent C_{19} cyclopropane fatty acid content of *Lactobacillus bulgaricus.* (From Smittle, R. B., Gilliland, S. E., Speck, M. L., and Walter, W. M., Jr., *Appl. Microbiol.,* 27, 738, 1974. With permission.)

unsaturated to saturated fatty acids (the C_{19} cyclopropane fatty acid was included as an unsaturated fatty acid).[54]

H. Antibiotics in Milk

The presence of residual antibiotics in milk is of great concern in the manufacture of all cultured dairy products. Most starter culture bacteria are sensitive to antibiotics. The lactobacilli are no exception in that they are sensitive to very low levels of antibiotics. This sensitivity of starter bacteria to antibiotics in milk stresses the importance of having a rapid method for testing milk for the presence of antibiotics prior to being used in the manufacture of cultured milk products.

I. Bacteriophage

As with the lactic streptococci and other bacterial species there are bacteriophage specific for the lactobacilli. While there have been few reports of bacteriophage for lactobacilli occurring in dairy products it is important that caution be used in handling the cultures to control bacteriophage.[55] Host specificity appears to exist for the phages for lactobacilli as is observed with other bacterial species.[56] Many of the lactobacilli exhibit lysogeny.[57-59] Within the streptobacterium group there is a very narrow range for host specificity that has been attributed to immunity caused by frequent lysogeny within this group.[59] In some species attempts to induce plaque formation from lysogenic lactobacilli through treatment of the cultures with mitomycin have resulted in the release of phage-like structures.[57,58] However, in several cases no plaques were formed, indicating a relatively low activity with regard to the lytic action of these types of bacteriophage. This lack of plaque formation could be due to insufficient knowledge of all the factors necessary to ensure proliferation of these phages. The bacteriophage for *L. casei* requires calcium ions for penetration into the host cell.[60] These phages also exhibit an optimum pH range of 5.5 to 6 and an optimum temperature of 30°C.

IV. BENEFITS OF LACTOBACILLI OTHER THAN ALTERING THE FORM OF MILK

Starter cultures such as the lactobacilli may improve the nutritive value of milk proteins by making them more digestible.[61-63] Certain of the lactobacilli are also important to human health from the standpoint of improving the digestion of lactose, particularly in those people who are considered lactose malabsorbers.[64,65] Additionally, *L. acidophilus* has been shown to be capable of reducing serum cholesterol levels in humans and more recently to have a role in retarding the formation of certain types of colon cancer.[66] Rats fed cultured yogurt grew better than those fed yogurt manufactured by a direct acidification process without the use of a starter culture.[63] In vitro tests showed that cultured yogurt was more digestible than uncultured mix from which the yogurt was made.[62] The free amino acid content of the cultured yogurt was twice that of the uninoculated mix.

Lactose malabsorbers (those persons who cannot adequately digest lactose) who consumed milk containing cells of *L. acidophilus* excreted less breath hydrogen, which is a measure of lactose malabsorption, than did those individuals who consumed milk without the lactobacilli.[64] Similar effects were observed for lactose malabsorbers consuming cultured yogurt compared to yogurt manufactured by direct acidification without the use of a starter culture.[65] There are quite a number of other reports in the literature dealing with the beneficial attributes of including lactobacilli or products fermented with lactobacilli in the diet. There have been several review articles in recent years that address this point.[66-69] It is important that future research is focused upon studying these possibilities and evaluating the mechanisms responsible for such benefits. If nutritional and health benefits are to be derived from such products it is very important that we understand the mechanism whereby the benefits are derived so that we may ensure that the culture maintains these desirable attributes during the production and manufacture of the culture-containing products.

REFERENCES

1. Kosikowski, F., *Cheese and Fermented Milk Foods*, 1st ed., Edwards Brothers, Ann Arbor, Mich., 1966.
2. Speck, M. L., Market outlook for acidophilus food products, *Cult. Dairy Prod. J.*, 10, 8, 1975.
3. Humphreys, C. L. and Plunkett, M., Yoghurt: a review of its manufacture, *Dairy Sci. Abstr.*, 31, 607, 1969.
4. Buchanan, R. E. and Gibbons, N. E., Eds., *Bergey's Manual of Determinative Bacteriology*, 8th ed., Williams & Wilkins, Baltimore, 1974.
5. Gilliland, S. E. and Speck, M. L., Use of the Minitek system for characterizing lactobacilli, *Appl. Environ. Microbiol.*, 33, 1289, 1977.

6. Rogosa, M. and Sharpe, M. E., An approach to the classification of the lactobacilli, *J. Appl. Bacteriol.*, 22, 329, 1959.
7. Wheater, D. M., The characteristics of *Lactobacillus acidophilus* and *Lactobacillus bulgaricus*, *J. Gen. Microbiol.*, 12, 123, 1955.
8. Sharpe, M. E., A serological classification of lactobacilli, *J. Gen. Microbiol.*, 12, 107, 1955.
9. Shimohashi, H. and Mutai, M., Specific antigens of *Lactobacillus acidophilus*, *J. Gen. Microbiol.*, 103, 337, 1977.
10. Shimohashi, H., Kodaira, H. S., and Suegara, N., Grouping antigens for four *Lactobacillus* species and their characteristics, *Jpn. J. Microbiol.*, 20, 405, 1976.
11. Knox, K. W., Campbell, L. K., Evans, J. D., and Wicken, A. J., Identification of the group G antigen of lactobacilli, *J. Gen. Microbiol.*, 119, 203, 1980.
12. Keenan, T. W. and Lindsay, R. C., Dehydrogenase activity of *Lactobacillus* species, *J. Dairy Sci.*, 50, 1585, 1967.
13. Keenan, T. W. and Lindsay, R. C., Diacetyl production and utilization by *Lactobacillus* species, *J. Dairy Sci.*, 51, 181, 1968.
14. Bottazzi, V. and Vescovo, M., Carbonyl compounds produced by yoghurt bacteria, *Neth. Milk Dairy J.*, 23, 71, 1969.
15. Lees, G. J. and Jago, J. R., Role of acetaldehyde in metabolism: a review. II. The metabolism of acetaldehyde in cultured dairy products, *J. Dairy Sci.*, 61, 1216, 1978.
16. Bautista, E. S., Dahiya, R. S., and Speck, M. L., Identification of compounds causing symbiotic growth of *Streptococcus thermophilus* and *Lactobacillus bulgaricus* in milk, *J. Dairy Res.*, 33, 299, 1966.
17. Argyle, P. J., Mathison, G. E., and Chandan, R. C., Production of cell-bound proteinase by *Lactobacillus bulgaricus* and its location in the bacterial cell, *J. Appl. Bacteriol.*, 41, 175, 1976.
18. Thomas, T. D. and Mills, O. E., Proteolytic enzymes of starter bacteria, *Neth. Milk Dairy J.*, 35, 255, 1981.
19. Chandan, R. C., Argyle, P. J., and Mathison, G. E., Action of *Lactobacillus bulgaricus* proteinase preparations on milk proteins, *J. Dairy Sci.*, 65, 1408, 1982.
20. El Soda, M., Bergere, J. L., and Desmazeaud, M. J., Detection and localization of peptide hydrolases in *Lactobacillus casei*, *J. Dairy Res.*, 45, 519, 1978.
21. Brandsaeter, E. and Nelson, F. E., Proteolysis by *Lactobacillus casei*. I. Proteinase activity, *J. Bacteriol.*, 72, 68, 1956.
22. Ohmiya, K. and Sato, Y., Studies on the proteolytic action of dairy lactic acid bacteria. VII. Action of intracellular protease of *Lactobacillus bulgaricus*, *Lactobacillus helveticus* or *Streptococcus lactis* on casein, *Agric. Biol. Chem.*, 33, 669, 1969.
23. McKay, L. L., Sandine, W. E., and Elliker, P. R., Lactose utilization by lactic acid bacteria: a review, *Dairy Sci. Abstr.*, 33, 493, 1971.
24. Lawrence, R. C. and Thomas, T. D., The fermentation of milk by lactic acid bacteria, in *Microbiological Technology: Current State, Future Prospects*, Bull, A. T., Ellwood, D. C., and Ratledge, C., Eds., Cambridge University Press, London, 1979, 187.
25. Premi, L., Sandine, W. E., and Elliker, P. R., Lactose-hydrolyzing enzymes of *Lactobacillus* species, *Appl. Microbiol.*, 24, 51, 1972.
26. Itoh, T., Ohhashi, M., Toba, T., and Adachi, S., Purification and properties of β-galactosidase from *Lactobacillus bulgaricus*, *Milchwissenschaft*, 35, 593, 1980.
27. O'Leary, V. S. and Woychik, J. H., Utilization of lactose, glucose, and galactose by a mixed culture of *Streptococcus thermophilus* and *Lactobacillus bulgaricus* in milk treated with lactase enzyme, *Appl. Environ. Microbiol.*, 32, 89, 1976.
28. Gunsalus, I. C., Horecker, B. L., and Wood, W. A., Pathways of carbohydrate metabolism in microorganisms, *Bacteriol. Rev.*, 19, 79, 1955.
29. Gunsalus, I. C. and Stanier, R. Y., *The Bacteria*, Vol. 2, *Metabolism*, Academic Press, New York, 1961.
30. Buyze, G., Van Den Hamer, C. J. A., and De Haan, P. G., Correlation between the hexosemonophosphate shunt, glycolytic system and fermentation type in lactobacilli, *Antonie van Leeuwenhoek; J. Microbiol. Serol.*, 23, 345, 1957.
31. Rutter, W. J. and Hansen, R. G., Lactose metabolism. II. The conversion of galactose to glucose derivatives in *Lactobacillus bulgaricus* strain Gere A, *J. Biol. Chem.*, 202, 323, 1953.
32. Auclair, J. E., The inhibition of microorganisms in raw milk. II. Distribution and properties of two inhibitory substances, lactenin 1 and lactenin 2, *J. Dairy Res.*, 21, 323, 1954.
33. Auclair, J. E. and Portman, A., Influence du chauffage du lait sur de develpment des bacteries. Effet stimulant du lait autoclave sur la croissance de *Lactobacillus lactis* (Role de l'acide formique), *Rev. Lait*, 39, 496, 1959.

34. Thompson, M. P. and Gyuricsek, D. M., Manufacture of yogurt, buttermilk, and cottage cheese from hydrolyzed lactose milks, *J. Dairy Sci.*, 57, 584, 1974.
35. Hemme, D., Vassal, L., Foyen, H., and Auclair, J., Effet de l'addition de lactase au lait sur le developpement des lactobacilles et des *Streptocoques thermophiles, Lait*, 59, 597, 1979.
36. Zuraw, E. A., Speck, M. L., Aurand, L. W., and Tove, S. B., Purification of stimulants from condensed corn-fermentation solubles active for *Lactobacillus casei* in milk, *J. Bacteriol.*, 80, 457, 1960.
37. Cogan, T. M., Gilliland, S. E., and Speck, M. L., Identification of stimulants for *Lactobacillus bulgaricus* in tomato juice, *Appl. Microbiol.*, 16, 1215, 1968.
38. Cogan, T. M., Gilliland, S. E., and Speck, M. L., Characterization of an inhibitor for *Lactobacillus bulgaricus* in tomato juice, *Appl. Microbiol.*, 16, 1220, 1968.
39. McAnelly, J. K. and Speck, M. L., Amino acid content of a peptide stimulatory for *Lactobacillus casei, J. Bacteriol.*, 73, 676, 1957.
40. Kitay, E. and Snell, E. E., Some additional nutritional requirements of certain lactic acid bacteria, *J. Bacteriol.*, 60, 49, 1950.
41. Marshall, V. M. E., Cole, W. M., and Mabbitt, L. A., Yoghurt made from single starter oganisms using heat or enzyme-treated milk or milk to which casein hydrolysate or sodium formate is added, *J. Dairy Res.*, 49, 147, 1982.
42. Thornhill, P. and Cogan, T. M., Effect of fruit on growth of *Lactobacillus bulgaricus* and *Streptococcus thermophilus, J. Dairy Res.*, 44, 155, 1977.
43. Gilliland, S. E. and Speck, M. L., Biological response of lactic streptococci and lactobacilli to catalase, *Appl. Microbiol.*, 17, 797, 1969.
44. Wheater, D., Hirsch, M. A., and Mattick, A. T. R., Possible identity of lactobacillin with hydrogen peroxide produced by lactobacilli, *Nature (London)*, 170, 623, 1952.
45. Gilliland, S. E. and Speck, M. L., Instability of *Lactobacillus acidophilus* in yogurt, *J. Dairy Sci.*, 60, 1394, 1977.
46. Archibald, F. S. and Fridovich, I., Manganese, superoxide dismutase, and oxygen tolerance in some lactic acid bacteria, *J. Bacteriol.*, 146, 928, 1981.
47. Accolas, J. P., Bloquel, R., Didienne, R., and Regnier, J., Acid producing properties of thermophilic lactic bacteria in relation to yoghurt manufacture, *Lait*, 57, 1, 1977.
48. Pette, J. W. and Lolkema, H., Yogurt. I. Symbiosis and antibiosis in mixed cultures of *Lactobacillus bulgaricus* and *Streptococcus thermophilus, Neth. Milk Dairy J.*, 4, 197, 1950.
49. Galesloot, T. E., Hassing, F., and Verenga, H. A., Symbosis in yogurt. I. Stimulation of *Lactobacillus bulgaricus* by a factor produced by *Streptococcus thermophilus, Neth. Milk Dairy J.*, 22, 50, 1968.
50. Driessen, F. M., Kingma, F., and Stadhouders, J., Evidence that *Lactobacillus bulgaricus* in yogurt is stimulated by carbon dioxide produced by *Streptococcus thermophilus, Neth. Milk Dairy J.*, 36, 135, 1982.
51. Moon, N. J. and Reinbold, G. W., Commensalism and competition in mixed cultures of *Lactobacillus bulgaricus* and *Streptococcus thermophilus, J. Milk Food Technol.*, 39, 337, 1976.
52. Smittle, R. B., Gilliland, S. E., and Speck, M. L., Death of *Lactobacillus bulgaricus* resulting from liquid nitrogen freezing, *Appl. Microbiol.*, 24, 551, 1972.
53. Smittle, R. B., Gilliland, S. E., Speck, M. L., and Walter, W. M., Jr., Relationship of cellular fatty acid composition to survival of *Lactobacillus bulgaricus* in liquid nitrogen, *Appl. Microbiol.*, 27, 738, 1974.
54. Goldberg, I. and Eschar, L., Stability of lactic acid bacteria to freezing as related to their fatty acid composition, *Appl. Envrion. Microbiol.*, 33, 489, 1977.
55. Peake, S. E. and Stanley, G., Partial characterization of a bacteriophage of *Lactobacillus bulgaricus* isolated from yoghurt, *J. Appl. Bacteriol.*, 44, 321, 1978.
56. Accolas, J. P. and Spillman, H., Morphology of bacteriophages of *Lactobacillus bulgaricus, L. lactis* and *L. helveticus, J. Appl. Bacteriol.*, 47, 309, 1979.
57. Yokokura, T., Kodaira, S., Ishiwa, H., and Sakurai, T., Lysogeny in lactobacilli, *J. Gen. Microbiol.*, 84, 277, 1974.
58. De Klerk, H. C. and Hugo, N., Phage-like structures from *Lactobacillus acidophilus, J. Gen. Virol.*, 8, 231, 1970.
59. Stetter, K. O., Evidence for frequent lysogeny in lactobacilli: temperate bacteriophage within the subgenus Streptobacterium, *J. Virol.*, 24, 685, 1978.
60. Watanabe, K. and Takasue, S., The requirement of calcium in infection with *Lactobacillus* phage, *J. Gen. Virol.*, 17, 19, 1972.
61. Speck, M. L. and Katz, R. S., ACDPI status paper nutritive and health values of cultured dairy foods, *Cult. Dairy Prod. J.*, 15, 10, 1980.

62. Breslaw, E. S. and Kleyn, D. H., In vitro digestibility of protein in yogurt at various stages of processing, *J. Food Sci.*, 38, 1016, 1973.
63. Hargrove, R. E. and Alford, J. A., Growth rate and feed efficiency of rats fed yogurt and other fermented milks, *J. Dairy Sci.*, 61, 11, 1978.
64. Kim, H. S. and Gilliland, S. E., *Lactobacillus acidophilus* as a dietary adjunct for milk to aid lactose digestion in humans, *J. Dairy Sci.*, 66, 959, 1983.
65. Gilliland, S. E. and Kim, H. S., Viable starter culture in yogurt aid lactose utilization in humans, *J. Dairy Sci.*, 67, 1, 1984.
66. Speck, M. L., Interactions among lactobacilli and man, *J. Dairy Sci.*, 59, 338, 1976.
67. Sandine, W. E., Muralidhara, K. S., Elliker, P. R., and England, D. C., Lactic acid bacteria in food and health: a review with special reference to enteropathogenic *Escherichia coli* as well as certain enteric diseases and their treatment with antibiotics and lactobacilli, *J. Milk Food Technol.*, 37, 691, 1972.
68. Sandine, W. E., Roles of lactobacillus in the intestinal tract, *J. Food Protect.*, 42, 259, 1979.
69. Gilliland, S. E., Beneficial interrelationships between certain microorganisms and humans: candidate microorganisms for use as dietary adjuncts, *J. Food Protect.*, 42, 164, 1979.

Chapter 5

THE LACTOBACILLI: MEAT PRODUCTS

James N. Bacus and William L. Brown

TABLE OF CONTENTS

I. HISTORICAL ASPECTS

Before the introduction and subsequent use of the strains of pediococci as the predominant meat starter cultures, most natural isolates from fermented meats consisted of various species and strains of lactobacilli.[1] Lactobacilli are still the predominant microflora in products that are "naturally fermented" from chance inoculation. An early U.S. patent pioneering the use of a bacterial starter culture in sausage products described the effectiveness of *Lactobacillus plantarum, L. brevis,* and *L. fermente.*[2] These investigators used several *Lactobacillus* species to effect the meat fermentation, but they depended upon chance inoculation for nitrate reduction (i.e., curing reaction). Subsequent work with pure cultures failed to achieve successful lyophilized preparations of *Lactobacillus* isolates with "state-of-the-art" technology, so continuing investigations focused on the *Pediococcus* strains that were resistant to the lyophilization process.[3]

With the introduction of frozen culture concentrates for sausage in 1968, interest in the *Lactobacillus* strains was renewed. In addition, the lactobacilli generally have lower growth temperatures that made them more suitable for dry sausage production. In 1974, a patent was issued that described the use of *L. plantarum,* either alone or in combination with other lactic acid producing microorganisms (i.e., *Pediococcus acidilactici*) for preparing dry and semi-dry sausages.[4] The concept of a "mixed" starter culture was also introduced to afford the sausage manufacturer a broad range of fermentation temperatures.

II. IDENTITY CHARACTERISTICS OF SPECIES INVOLVED

L. plantarum (NRRL-B-5461) is described as Gram-positive, nonmotile rods, ordinarily 0.6 to 0.8 μm × 1.2 to 6 μm, occurring singly or in short chains. The organism is catalase negative and does not grow at either 7 or 45°C. The thermal death point is 30 min at 63°C. Glucose fermentation results in the production of DL-lactic acid with no gas, and the strain is particularly salt tolerant, developing in salt concentrations greater than 9%. A variety of carbohydrates are fermented including fructose, glucose, galactose, sucrose, maltose, lactose, dextrin, sorbitol, mannitol, and glycerol. Xylose, dulcitol, mannose, and salicin are among those carbohydrates not fermented. The mol % G + C content has been reported as 43.6 compared to the 45 ± 1 mol % described for *L. plantarum* in *Bergey's Manual of Determinative Bacteriology.*[5]

L. plantarum is one of 27 described species in the genus *Lactobacillus* (Genus I) that is within the Family (I) Lactobacillaceae: Gram-positive, asporogenous, rod-shaped bacteria. *L. plantarum* is generally differentiated from many of the other species by the lack of gas production from glucose, but gas is produced from gluconate. Ribose is fermented yielding lactic and acetic acids, thiamin is not required, aldolase activity, glucose 6-phosphate dehydrogenase, and inducible 6-phosphogluconate dehydrogenase are present. The lactic fermentation produces DL-lactic acid.

The organism is a facultative anaerobe and typically does not reduce nitrate to nitrite. Ammonia is not produced from arginine and milk is acidified. Optimal growth is usually at 30 to 35°C. Calcium pantothenate and niacin are required. Natural isolates are obtained from dairy products and environments, fermenting plants, silage, sauerkraut, pickled vegetables, and the human intestinal tract. Closely related species are *L. casei* subsp. *pseudoplantarum* and *L. curvatus.*[5]

Natural isolates described as lactobacilli have been repeatedly recovered as the predominant microflora from fermented meat and sausage products.[1] Most authors have emphasized the occurrence of *L. plantarum,* but descriptions of *L. casei* and *L. leichmanii* have also been documented.[6] In general, the majority of the natural isolates are

Table 1
FERMENTATION RATES OF TWO STRAINS OF LACTOBACILLI[4]

	pH of the sausages aged at 65°F (18.3°C)		
Days	Control	NRRL-B-5461	ATCC-14917
1	5.80	5.75	5.75
2	5.80	5.68	5.75
3	5.55	5.20	5.45
4	5.22	4.60	4.85
	pH of the sausages aged at 100°F (37.8°C)		
Hr	Control	NRRL-B-5461	ATCC-14917
0	5.80	5.80	5.80
7	5.83	5.80	5.85
17	5.88	5.13	5.75
21	—	4.60	5.20
24	5.85	4.45	5.10
41	—	4.45	4.53

species variants of lactobacilli that do not precisely conform to any specific species, but they are most closely related to *L. plantarum*. These strains are highly adapted to the specific environment from which they were isolated, and they will out-perform classical strains when both are reintroduced into the sausage environment. A natural isolate, *L. plantarum* NRRL-B-5461, when compared to a known strain (ATCC-14917) in the fermentation of a typical sausage mix developed a lower pH in a shorter time at two different fermentation temperatures (Table 1).[4]

Most commercial starter cultures composed of lactobacilli are homofermentative, yielding primarily lactic acid from the fermentation of dextrose and sucrose. However, heterofermentative lactobacilli can often occur as natural contaminants and contribute to the sausage flavor through the production of volatile acids, alcohol, and carbon dioxide. The role of two heterofermentative species, *L. brevis* and *L. buchneri*, has been documented and the incorporation of heterofermentative lactics in sausage-making has been suggested.[7]

The wide variety of strains of lactobacilli probably contributes to the "uniqueness" of many dry sausage products. Comparisons of commercial *L. plantarum* starter cultures, classical ATCC strains, and various natural isolates of lactobacilli from sausage demonstrate variations in the respective fermentation characteristics, growth temperatures, and the rate and type of acid production (Table 2).[8] No appreciable proteolytic or lipolytic activity was detected in any of the strains. All strains demonstrated the ability to decompose hydrogen peroxide when grown in the presence of myoglobin. This "catalase activity" would retard any oxidation and the destruction of meat pigments that can occur when peroxides are present. Catalase activity has been promoted as a major benefit when employing micrococci as meat starter cultures, and it would appear that lactobacilli would also have this activity when propagated in meat.[8] It has been observed that some strains of lactobacilli used as meat starters are unable to produce and accumulate hydrogen peroxide,[9] although this characteristic is highly variable, and the observed results are very dependent on analytical technique.

Table 2
BIOCHEMICAL CHARACTERISTICS OF LACTOBACILLI USED IN DRY SAUSAGE PRODUCTION[a] [e]

	Substrate (API 50L System)[b]																		Growth temp. (°C)				
Strain	Glycerol	L-(+)-arabinose	Mannitol	Sorbitol	Methyl-D-glucoside	Amygdalin	Arbutin-iron-citrate	Esculin-iron-citrate	Salicin	D-(+)-cellobiose	Maltose	Lactose	D-(+)-melezitose	D-(+)-raffinose	Arginine	Glucose, gas production	Teepol, 9.4%	Teepol, 0.6%	o-Nitrophenyl-β-D-galactosidase	42	10	8	Rel. catalase activity[c]	Final pH[d]
ATCC 8014	±[e]	+	+	+	+	+	+	+	+	+	+	+	+	+	−	±	+	+	+	+	+	−	100	4.5
SL	+	+	+	+	+	+	+	+	+	+	+	+	−	+	−	+	+	−	+	+	−	−	121	4.4
DL	−	−	+	+	−	+	+	+	+	+	+	+	+	+	−	+	+	+	+	+	+	−	35	4.8
I-L	−	−	−	−	−	+	±	+	−	−	+	−	−	−	−	±	−	−	±	+	+	+	76	4.9
II-L	−	−	−	−	−	−	−	+	−	+	−	−	−	−	±	±	−	−	−	−	+	+	96	4.6
III-L	−	−	−	−	−	−	−	−	−	−	−	−	−	−	−	±	−	−	−	−	+	+	119	5.3
IV-L	−	+	−	−	−	−	±	+	+	±	+	−	−	−	+	±	−	−	+	−	+	+	92	4.8
V-L	+	+	−	−	−	−	−	±	−	±	±	±	−	−	±	±	−	−	+	−	+	+	156	5.0

[a] The following reactions were positive for all the strains: ribose, galactose, D-(+)-glucose, D-(−)-levulose, fructose, D-(+)-mannose, N-acetylglucosamine, D-(+)-melibiose, sucrose, D-(+)-trehalose, and growth at 13 and 40°C. The following reactions were negative for all the strains: erythritol, D-(−)-arabinose, D-(+)-xylose, L-(−)-xylose, adonitol, methylxyloside, L-(−)-sorbate, rhamnose, dulcitol, meso-inositol, methyl-D-mannoside, inulin, dextrin, amylose, starch, glycogen, acidification of gluconate, urease, catalase, nitrate reduction, acetoin production, and growth at 45°C.

[b] API system, S. A. LaBalme les Grottes, Montalieu Vercieu, France.

[c] The activity of strain ATCC 8014 was used as reference.

[d] M_3 medium containing 1.8% glucose, initial pH 6.9, 3-day incubation at 25°C.

[e] ± Variable.

III. FUNCTIONS OF LACTOBACILLI IN MEAT PRODUCTS

Some meat products, specifically certain dry and semi-dry sausages, undergo a microbial fermentation at some phase in their processing cycle. This natural fermentation evolved with the development of these types of sausages since the responsible microorganisms were associated with the meat and processing environment, and the formulation (i.e., salt, nitrite, sugar) and processing parameters (i.e., smoke, anaerobic environment, low temperature aging) favored their subsequent growth and metabolism. The fermentation reduced the meat pH, thereby preserving the product against undesirable microorganisms and contributing to the palatability characteristics. The lower meat pH also enhanced the drying process that was characteristic in the preparation of these sausages.

Recently, major manufacturers of fermented sausage products in the U.S. have proposed guidelines for good manufacturing practices for dry and semi-dry sausages.[10] They have defined dry sausages as "chopped or ground meat products that, as a result of bacterial action, reach a pH of 5.3 or less and are then dried to remove 20 to 50% of the moisture to have a moisture to protein ratio no greater than 2.3 to 1.0". Semi-dry, fermented sausages are defined as "chopped or ground meat products that, as a result of bacterial action, reach a pH of 5.3 or less and undergo up to 15% removal of moisture during the heating process. In general, the semi-dry sausages are not subsequently dried in a dry room but they are packaged soon after the fermentation/heating process is completed. They are generally smoked during the fermentation cycle and have moisture to protein ratios no greater than 3.7 to 1.0". The proposed guidelines additionally state that this pH reduction is attributed to the action of lactic acid-forming bacteria that may be added to the meat by either a commercially prepared starter culture or a "back inoculum" from a previously fermented "mother batch".

Meat starter cultures composed of lactobacilli are most often utilized in the production of dry sausages where fermentation temperatures are between 15 and 35°C. In these types of products, the primary lactic acid fermentation is necessary to achieve the flavor and stability characteristics of the product. In addition, the resulting lower pH allows the meat to release moisture more readily and uniformly, and the acid denatures the meat protein, contributing to the characteristic, firm texture. Secondary fermentations and other chemical reactions often play a significant role in the organoleptic distinctiveness of these products. Fermentation times range from 1 to 5 days in the more traditional dry sausage processes. These lower temperature fermentations for extended times may allow some growth and metabolism by the naturally occurring microflora resulting in various metabolic end products that affect the final flavor. Enzymatic reactions in the meat also have more time to be effected prior to cooking, drying, and/or pH reduction. Most dry sausage products do not have the sharp, "tangy" taste that is characteristic of semi-dry sausages. This is especially true with European products where a "mild" tast is preferred, and lower final product acidity (i.e., pH 5.3 to 5.6) are more frequently observed.[11]

In addition to their acid-producing capabilities, cultures of lactobacilli function as effective preservatives against the proliferation of food pathogens and other undesirable microorganisms. *Staphylococcus aureus* is a common pathogen associated with salted meat products and it can be inhibited by cultures of lactobacilli (Table 3).[12] Staphylococcal food poisoning caused by defective, fermented dry or semi-dry sausage is a potentially significant problem according to a USDA Task Force Report.[13] Since 1967, there have been at least six publicized incidents of staphylococcal food poisoning traced to fermented sausage.[13,14] These incidents involved different major meat companies and large amounts of product. In each case, there was a lack of scientifically based controls designed to ensure that the final product was safe for consumption.

Table 3
STAPHYLOCOCCAL ENTEROTOXIN DEVELOPMENT, DRY SAUSAGE, 22 to 24°C[12]

	3 Days			7 Days		
Starter culture	Log CPS[a]	pH	Enterotoxin	Log CPS	pH	Enterotoxin
−	8.84	5.9	+	8.88	5.7	+
+	6.78	5.6	−	7.53	5.3	−

[a] Log coagulase positive staphylococci per gram.

Table 4
TIME-TEMPERATURE CONTROL, DRY SAUSAGE[10]

Constant temp. processes; max. degree/hr to reach pH 5.3[a]	Temp. (°C)	Allowed hr by guideline
1200	23.9	80
1200	26.7	60
1200	29.4	48
1000	32.2	33
1000	35.0	28
1000	37.8	25
900	40.4	20
900	43.0	18

[a] Degrees are measured as excess over 15.6°C. Degree/hr are the product of time in hours at a particular temperature and the "degrees".

Further, the Task Force concluded that the current trends in production practices with increased volumes per facility and reduced processing times at higher temperatures may enhance the staphylococcal problem. Specific recommendations, in addition to a continued emphasis on good raw materials and proper sanitation, include controlled acidulation of all fermented sausage with either microbial starter cultures or chemical acidulants. The National Academy of Sciences has also recommended that manufacturers add lactic starter cultures and/or chemical acidulants to inhibit staphylococcal multiplication.[15]

Proper control mechanisms in the successful production of fermented sausage must be applied to the initial formulation and processing stages.[13] The Good Manufacturing Practices (GMPs) for fermented dry and semi-dry sausage particularly address those critical control points so as to minimize the opportunity of *S. aureus* reaching levels of public health significance. They state that "once the sausage pH reaches 5.3 or less, the environment for *S. aureus* is effectively controlled, thus minimizing the potential for growth to a dangerous level. During fermentation of sausages to a pH of 5.3, it is necessary to limit the time which the sausage meat is exposed to temperatures exceeding 15.6°C or higher." In addition to encouraging manufacturers to use controlled acidification procedures (mainly starter cultures), the GMPs also outline time-temperature-pH limits "which the industry has found to be safe, acceptable and attainable under practical production conditions" (Table 4).[10]

Although the acid production by the lactobacilli is the primary mechanism for inhibition, some researchers have demonstrated inhibitory characteristics by cultures of

Table 5
BOTULINAL TOXIN DEVELOPMENT, SUMMER-STYLE SAUSAGE AT 27°C[20]

Nitrite (ppm)	Starter culture	Dextrose	Toxic/25
0	−	+	8
0	+	−	22
0	+	+	2
50	+	+	0
150	−	−	14
150	+	+	0

Table 6
BACON PROCESSING — AVERAGE ANALYSES — 21 DAYS[a]

	Residual nitrite (ppm)	Standard plate count	pH	Nitrosopyrrolidine (ppb)
Control	20—40	10^4—10^5	6.0—6.4	10—30
Lactobacillus blend	4—16	10^6—10^7	5.2—5.6	2—9

[a] Typical range of values observed.

lactobacilli in nonfermented meats such as ham, bacon, and ground beef.[16-18] Presumably, this inhibition is due to the production of hydrogen peroxide and/or other substances, as antibiotics, bacteriocins, etc.[18] Meat starter cultures composed of lactobacilli also have demonstrated inhibitory effects in sausages against *Salmonella*[19] and *Clostridium botulinum* (Table 5).[20]

Through the control of undesirable microorganisms, *Lactobacillus* cultures also serve as an effective control agent against the production of histamines (a common toxicant associated with fermented foods) by naturally occurring microbial contaminants that possess histidine decarboxylase activity.[21] Similar results have demonstrated that the use of lactobacilli decreases the possibility of potentially hazardous tyramine concentrations that can result from a microflora that possess both tyrosine decarboxylase and proteolytic activities.[22] Several reports have confirmed the lack of histamine and tyramine production by commercially available *Lactobacillus* meat starter cultures.[23]

Lactobacilli in combination with a fermentable substrate are an effective antibotulinal agent in fermented meats.[20] The lower pH, attained through fermentation, also dissipates residual nitrite concentration by accelerating its conversion to nitrous oxide.[24] This natural control mechanism has been applied successfully to bacon processing (to minimize nitrosamine formation) through the addition of a *Lactobacillus* starter culture to the curing pickle.[17] During smoking and/or subsequent storage the lactobacilli reduce product pH, achieving product stability, yet lowering residual nitrite. Both factors reduce potential nitrosamine formation at the time of frying (Table 6). Residual nitrite can react with secondary amines in certain meat products, particularly bacon fried at high temperatures to form nitrosamines which are proven carcinogens.[25] Reducing the residual nitrite content and/or reducing the product pH will decrease nitrosamine formation. Since bacon traditionally is not cooked in excess of 55°C, the culture is also able to survive most processing schedules and provide effective protection during any subsequent temperature abuse. Reducing the nitrite will minimize nitrosamine formation, but it will also increase the potential for botulinal toxin formation (Table 7).[26] The cultured product actually becomes "safer" with prolonged abuse since

Table 7
BOTULINAL TOXIN
DEVELOPMENT, BACON AT 27°C[26]

Nitrite	Sucrose	Lactobacilli	Toxic/samples
0	–	+	26/27
0	+	–	50/52
0	+	+	1/49
40	–	–	47/50
40	+	+	0/30
120	–	–	17/28
120	+	+	1/192

the lactobacilli serve as a "safety valve" that readily activates with increased temperature. The increased metabolic activity rapidly lowers meat pH, thereby inhibiting any *C. botulinum* or other food pathogens. Quality attributes of "cultured" bacon are not normally affected with proper handling and storage conditions.

Since the safety and suitability of these microorganisms had been previously established for food products in the Federal Meat Inspection Act (9CFR318.7) and by the Food and Drug Administration (GRAS), lactic acid starter cultures have been officially USDA-approved for use in bacon to lower or eliminate nitrosamines.[27] A full year of commercial bacon production substantiated the ability of the process to minimize nitrosamine formation while maintaining the integrity of the product.[28] This effective system may explain the relative lack of nitrosamine formation in fermented sausage,[29] and it may offer a viable alternative in dry-cured meats, where nitrosamines have been a problem.[30]

Fermented meat products traditionally have demonstrated an extended shelf-life through a combination of reduced moisture content and pH. The USDA recognizes sausage having a moisture/protein ratio of 3.1 or less, and a pH of 5.0 or less, as not requiring refrigeration.[13] Additionally, shelf-stable meat products are classified as having a pH at or below 5.2 and a water activity at or below 0.95, or a pH below 5.0 or a water activity below 0.91.[31] The shelf-life of these products is generally not limited by bacterial deterioration, but by chemical or physical spoilage. Microbial cultures contribute to the shelf-life of fermented meats mainly by consistent and controlled acidification. The acid development inhibits undesirable microorganisms and allows efficient dehydration.

IV. FACTORS INFLUENCING OPTIMUM PERFORMANCE

As the technology of meat fermentation has increased, the responsible lactic acid microorganisms have been isolated and reintroduced as pure starter cultures for the manufacture of these fermented meat products. Through the expanded use of starter cultures, the initial type and number of microorganisms inoculated into a specific meat product has become more consistent. As a result, other environmental factors affecting the meat fermentation have become more apparent. Although these formulation and process parameters affect both "wild" and "cultured" fermentations, the constant inoculum afforded by the use of a starter culture has allowed the other environmental factors to be more readily observed.

The rate of fermentation and the ultimate pH of the meat product are directly influenced by the specific formulation and processing conditions, as well as the type and "activity" of the culture employed. Since the safety and quality of the product are dependent on the rate and extent of acid production, a thorough understanding of these environmental parameters is essential to the total control of the product.

A. Meat

Although the meat comprises the majority of a sausage formulation, it is often overlooked as a factor affecting fermentation. The lactic microorganisms operate in the water phase of the formulation, thus any factor affecting the amount of available water will influence the microbial activity. The greater the moisture content of the meat, the greater the fermentation rate. The fat content is also important as it is inversely proportional to the lean meat percentage. In general, the higher the lean percentage (i.e., lower fat), the greater the moisture, resulting in a more rapid pH drop during fermentation. The contribution of glycogen from lean tissue has an apparent effect on total acidity.[32] The use of freeze-dried meat as an ingredient in dry sausage will retard the initial fermentation rate due to the reduced moisture.[33]

The buffering capacity (i.e., the ability of the meat to absorb acid) will also affect the rate of pH drop. The greater the buffering capacity, the more acid that must be produced by the bacteria prior to lowering product pH. This results in a slower overall fermentation time. The initial pH of the meat is also important as to subsequent fermentation time and final product pH. Meats with higher pH values will require more acid production to achieve the same end point.[34]

The ratio of fresh vs. frozen meat will influence the fermentation in several ways. Since microorganisms have an optimum growth temperature, the internal *product* temperature is critical to the fermentation time. A formulation high in frozen materials will take longer to achieve the desired fermentation, thereby extending the total incubation period. In addition, frozen meats are often somewhat dehydrated and exhibit rapid drip loss during thawing. A reduced moisture level attributed to previously frozen raw materials will reduce the rate of fermentation.

Many processors note that pork and beef products ferment at a faster rate than their all-beef counterparts. Several hypotheses include higher lactic acid bacterial contamination levels found in pork and/or the higher thiamin concentrations. In addition, beef will generally have a higher initial pH and greater buffering capacity.

The microbial flora of the raw meat materials attributed to prior handling conditions and sanitation may also influence subsequent fermentation through either direct microbial interactions and/or indirect chemical changes in the raw materials. High levels of lactic acid microorganisms may actually increase the rate of fermentation while undesirable microorganisms (*Pseudomonas,* yeast, etc.) may produce end products, either prior to, or during the lactic fermentation, that affect the flavor. High initial numbers of certain yeasts can compete successfully with the lactic acid microorganisms and retard pH drop through the production of more basic end products, including alcohol. Since most yeasts are acid tolerant, this can also occur after the fermentation has been completed, resulting in a pH increase and the development of off-flavors during drying. A high oxygen content in the meat may induce a microbial oxidation of a portion of the available carbohydrate with the production of carbon dioxide, water, alcohols, and carbonyl compounds.[35] Micrococci present in the early stages of fermentation may also be responsible for the complete conversion of carbohydrate. Either of these situations can result in a higher pH than expected. Many of the flavor differences and problems in fermented sausages can be attributed to the differences in how the added carbohydrates are metabolized by various groups of bacteria that predominate in a unique process.[34]

B. Nonmeat Ingredients

Salt is added to fermented meat products to achieve the desired bind (i.e., myosin extraction), flavor, and preservative qualities. Salt is the major component that allows the lactic acid bacteria to predominate and inhibits many undesirable microorganisms.

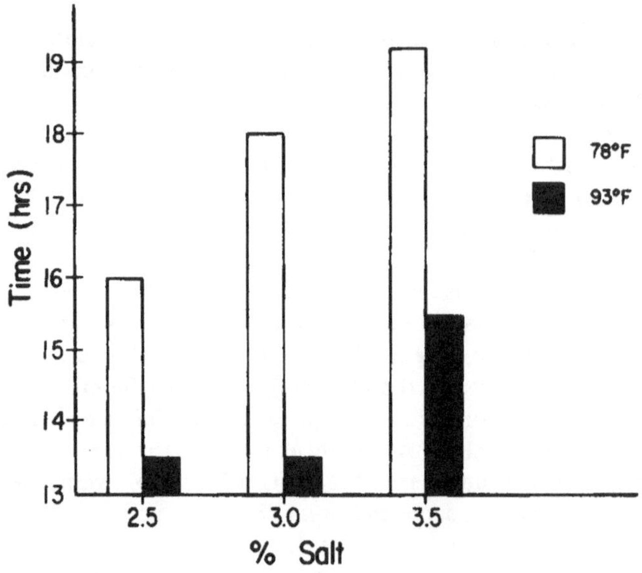

FIGURE 1. The effects of salt concentration and temperature on the sausage fermentation time of a blend of lactobacilli to reach pH 5.1.

In general, fermented sausages are formulated with 2.0 to 3.5% salt, depending on the nature of the product. Although the lactic microorganisms responsible for the fermentation are salt tolerant, the brine concentration (percent salt/percent moisture) of the formulation directly affects their performance. A salt level of 2% is regarded as a minimum to achieve the desired bind, and no major differences in fermentation rate are generally observed up to 3%.[36] However, salt concentrations in excess of 3% (i.e., 5 to 6% brine) will begin to lengthen fermentation time (Figure 1). Since salt is commonly employed as a "carrier" for curing agents and flavorings, the calculated salt content of the formulation should also include this portion.

Although salt definitely contributes to the stability of fermented meats, a reduced pH is also essential. An excess of salt in the initial formulation will not compensate for an extended fermentation time. Food poisoning staphylococci, which are a primary concern in fermented products, are more salt resistant than most lactic acid bacteria and therefore, can better tolerate increasing salt concentrations. A sausage formulation containing 4% salt may actually favor the growth of many contaminating staphylococci over the lactic acid microorganisms.

Various sugars (dextrose, sucrose, corn syrups) are added to the formulation to achieve the desired flavor, texture, and yield characteristics. They also provide the necessary fermentation substrates for the production of lactic acid. The type and amount of sugar can directly affect the ultimate pH of the product. Simple sugars, such as dextrose, are readily utilized by all lactic acid bacteria and their availability tends to be the limiting factor for sausage fermentation. It has been noted by some researchers that 1% fermentable carbohydrate will generally effect a decrease of 1 pH unit. Where the initial pH is 6.0, 1% dextrose should be added to promote an adequate drop in pH.[37] If the fermentation is allowed to continue to completion, the final pH is directly proportional to the initial dextrose level, up to approximately 0.5% (Figure 2). It has been generally recommended that sausage mixtures contain a minimum of 0.75% dextrose to afford a surplus of a known, fermentable carbohydrate. Cane sugar or sucrose can also be utilized by most lactic bacteria and will generally result in a "less

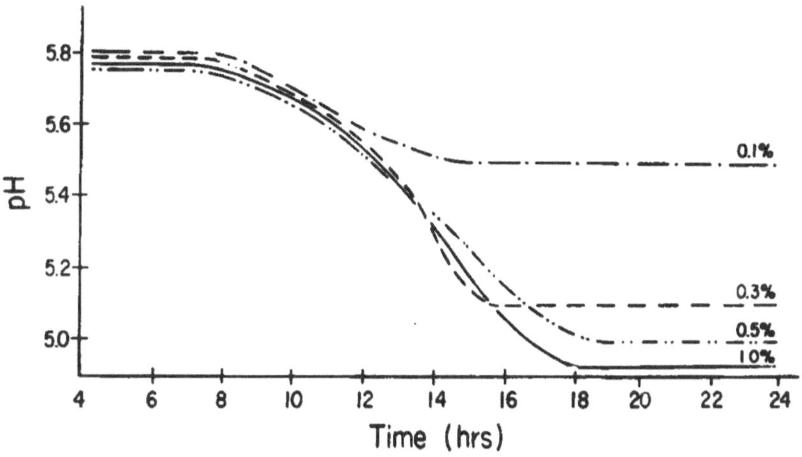

FIGURE 2. The effects of varying initial dextrose concentration on sausage fermentation by a blend of lactobacilli.

acid tasting" product when compared to dextrose at an equivalent pH. This probably results from either the greater "sweetness" from the residual sugar, or the utilization of either the fructose/dextrose component to yield a more basic end product (i.e., dextran, levan.)

More complex carbohydrates such as corn syrup, dextrin, flour, and starches may be fermented to various degrees depending upon their availability and the specific culture employed. These carbohydrates ferment more slowly and are not of practical significance where simple sugars are present, except in some dry sausages where unique flavors are produced over extended drying periods. The amount of acidity obtained from corn syrups was found dependent on the quantity of the simple carbohydrates, dextrose and maltose, initially available (i.e., dextrose equivalent.)[37] Excess concentrations of added carbohydrates (>2%) also can reduce the rate of fermentation through the binding of the available water.

Certain natural spices typically used in the formulation can have a direct effect on the rate of fermentation by stimulating acid production in the bacteria (Figure 3).[38] Often, the same processor with various products differing in spice composition observes differing rates of fermentation. Generally, this stimulation is not accompanied by an increase in the bacterial population and does not result from any contaminating microflora in the spices. Black, red, and white pepper, mustard, garlic powder, allspice, nutmeg, ginger, mace, and cinnamon all have been shown to stimulate acid production to varying degrees depending on the concentration and the culture employed. Generally, lactobacilli are more stimulated than the pediococci. Combinations of certain spices can often yield a shorter fermentation time than individual spices alone. The degree of acid stimulation can depend on the origin and type of the particular spice component (i.e., Lampong black pepper vs. Brazil black pepper). Recent studies have identified manganese as the factor in spices responsible for the enhancement of the acid production.[39] Stimulatory activity of acid extracts of spices increased with increasing manganese concentrations. Fermented sausages without spices but with added manganese (10^{-5} *M*) developed a similar level of activity to those sausages with added spices. This finding correlates to similar results documented in an early U.S. patent covering meat starter cultures.[40] In addition to an increased rate of acid production, the ultimate pH of the product is often lower when either natural spices or manganese is added to the sausage mix.

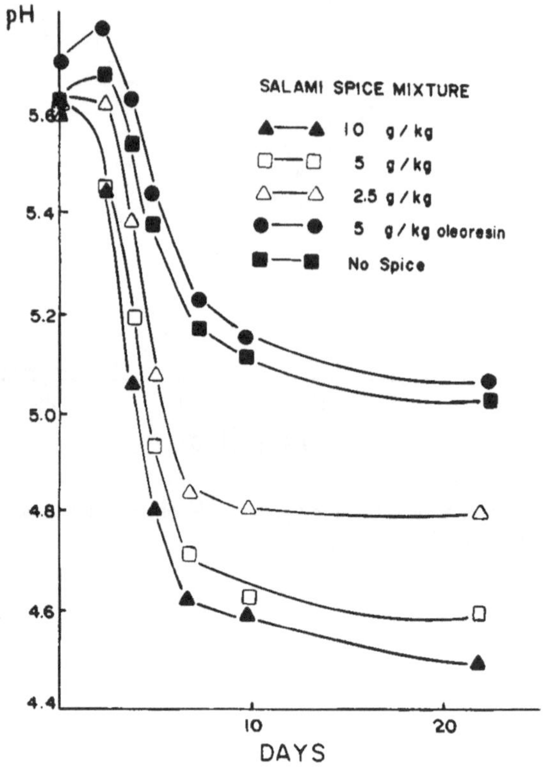

FIGURE 3. pH-course of dry sausage during fermentation in the presence of varying spice mixtures. (Reprinted from *J. Food Sci.*, 1982, 47, 1618. Copyright © by Institute of Food Technologists. With permission.)

Inhibiting effects have been observed with some natural spices, and particularly with the extracts of some spices (i.e., pepper), generally the volatile fraction. Some results have suggested that the normal levels of pepper used in sausages are not adequate to inhibit bacterial growth unless it is used as the essential oil.[41] This probably accounts for the somewhat longer fermentation times generally observed with liquid spice blends compared to their natural spice equivalents. In view of recent findings, if this is a true inhibition or just a lack of stimulation by the manganese content of the natural spices remains to be seen.

Other nonmeat ingredients typically encountered in formulations for fermented sausage can exert different effects on the microbial fermentation. The practical significance attributed to each component is dependent on the concentration, the culture activity, the remaining formulation, and the processing conditions. Generally, liquid smoke flavorings and antioxidants retard the fermentation rate.[42,43] Phosphate will function as a buffer, increasing the "lag time" prior to observing a pH decrease. Milk powder, soy proteins, and other dry powders can slow the fermentation time due to binding the available water. Formulations with added water will ferment at a faster rate. Sausages containing sodium nitrite will generally ferment slower than those without nitrite, but the degree of difference is largely dependent on the specific culture strain.

C. Process Procedures

The processing conditions affect both the rate of fermentation and the ultimate pH

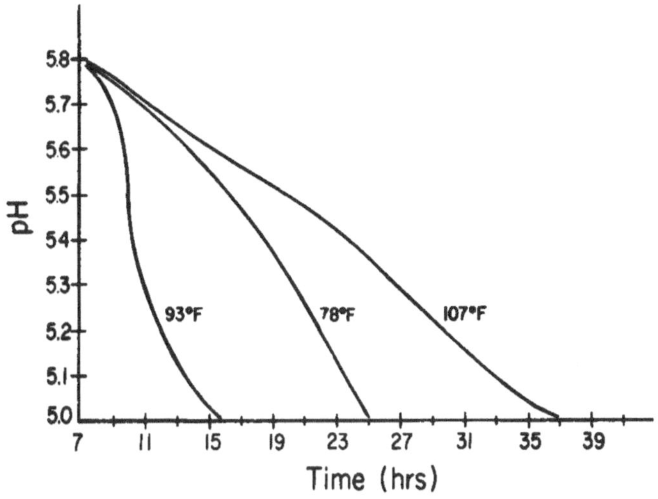

FIGURE 4. The effect of temperature on sausage fermentation by a blend of lactobacilli.

of the product. In the "wild" fermentations, the processing conditions select the type of microorganisms that will predominate which, in turn, determine the characteristics of the fermentation and the final flavor of the product. Processing parameters such as time, temperature, humidity, and smoke can be utilized effectively to control the total process and thus prevent problems.

Product temperature largely will determine the metabolic activity of the microorganisms present (Figure 4). The lactobacilli employed as starter cultures have an optimum growth temperature of 32°C. Their initial performance in lowering pH will be dictated by the product temperature and the time held at that temperature. Slower fermentation rates will be observed as the product temperature deviates from the optimum. In many instances, slower fermentations at lower temperatures are more desirable in controlling ultimate pH and developing flavor, color, and other product characteristics. Lower temperatures are also utilized to control undesirable pathogenic microorganisms. Higher finishing temperatures are employed in many products to stop the fermentation phase at a desired pH. Although internal product temperatures of 63 to 68°C usually are required to effectively kill the lactic acid microorganisms, temperatures of 46 to 52°C for extended periods can be employed to halt the acid production by the lactobacilli. An important factor to be considered in accounting for fermentation time is the "come up" time to achieve the desired product temperature. This is directly influenced by the initial product temperature, the fermentation temperature, the load in the house, air circulation, and humidity. Some processors observe seasonal variations in fermentation time with basically the same formulation and process.

Higher humidity favors more rapid fermentation since any drying of the product reduces available water. Dry sausages held with static air conditions usually ferment faster, and subsequently dry better, than those fermented with high air circulation. Although the latter conditions will effect more rapid heating, the surface of the meat will tend to dry out, irreversibly sealing the surface pores and preventing uniform drying.

Smoke is definitely inhibitory to microorganisms. In a larger diameter product, smoke will not exert a major effect on fermentation since the smoke penetration is limited to the surface. However, in the smaller diameter product (13 to 18 mm), heavy smoke application at the beginning of the fermentation cycle may retard the microbial activity. Smoke also may have a tendency to dry the product.

Casing diameter is important in predicting fermentation time and controlling ultimate pH in that larger diameter products generally exhibit lower ultimate pHs than the identical formulation processed in smaller casings. Even though the fermentation may initially proceed slower in the large diameter product due to slower heat penetration, the fermentation is also more difficult to stop with subsequent heat treatment or drying.

In dry sausages, the ultimate pH has been attributed mainly to the buffering capacity of the proteins and to the concentrations of ammonia and lactic acid. A significant negative correlation was found between moisture content and pH,[44] possibly related to the degree of lactic acid dissociation. This may explain the gradual increase in pH in many dry sausages during the aging process and the increase in pH of some semi-dry products when fully cooked.

Formulation and process uniformity determines fermentation consistency. Uniform distribution of formulation ingredients is essential to maintain a constant microenvironment within each batch and from batch to batch. Nonuniform distribution of salt, cure, sugar, spices, and/or starter culture may yield variations in the rate of fermentation and ultimate pH from sausage to sausage. This can result in flavor, texture, and stability problems in those pieces not adequately fermented. Variation also can result from product "positioning" within the smokehouse and dry room. Areas of higher temperature and humidity will generally ferment faster to a lower pH.

Although not a common occurrence, process contaminants may retard the fermentation rate. Boiler treatment compounds, when improperly used, can be toxic to the lactic microorganisms if exposed through a high humidity process. Sanitizing agents definitely are lethal to starter cultures and can drastically reduce culture activity through direct contact. Containers used to dilute and distribute the starter culture should be rinsed thoroughly after sanitizing. Additionally processing equipment should be rinsed so as to eliminate residual sanitizers prior to formulation.

D. Culture

The type and activity of the microbial starter culture or "seed" inoculum is a key element in structuring a consistent fermentation process. Knowledge of the optimum growth temperature and the effects of various additives will provide insight into any product variation and problems. Optimum storage and handling conditions for the culture are critical in maintaining consistent performance, as is uniform distribution throughout the batch. The distribution medium, usually tap water, should be periodically monitored for excessive levels of chlorine or metals if culture activity appears to decrease. The culture should never be directly mixed with the cure, salt, or other dry ingredients prior to blending into the meat.

Meat fermentation is a biological acidulation whereby unique product characteristics result from a microbial lowering of the pH of meat. The biological nature of the process makes it susceptible to many environmental factors that must be controlled to yield a consistent product. The use of a commercial starter culture provides an additional control mechanism through a consistent inoculum; however, it will not replace a thorough knowledge and control of the total process. Both "cultured" and "wild" fermentations involve similar lactic acid microorganisms whose function must be properly understood to achieve the desired result.

REFERENCES

1. Deibel, R. H., Niven, C. F., and Wilson, D. D., Microbiology of meat curing. III. Some microbiological and related technological aspects in the manufacture of fermented sausages, *Appl Microbiol.*, 9, 156, 1961.
2. Jensen, L. B. and Paddock, L. S., Sausage Treatment, U.S. Patent 2,225,783, 1940.
3. Deibel, R. H. and Niven, C. F., *Pediococcus cerevisiae:* a starter culture for summer sausage, *Bacteriol. Proc.*, 1957, 14, 1957.
4. Everson, C. W., Danner, W. E., and Hammes, P. A., Process for Curing Dry and Semi-Dry Sausages, U.S. Patent 3,814,817, 1974.
5. Buchanan, R. E. and Gibbons, N. E., Eds., *Bergey's Manual of Determinative Bacteriology*, 8th ed., Williams & Wilkins, Baltimore, 1974.
6. Anon., Some aspects of dry sausage manufacture, *Fleischwirtschaft*, 5, 748, 1978.
7. Urbaniak, L. and Pezacki, W., Die Milschsaure bildende Rohwurst-Mikroflora and ihre technologisch bedingte Veranderung, *Fleischwissenschaft*, 55, 229, 1975.
8. Nordal, J. and Slinde, E., Characteristics of some lactic and bacteria used as starter cultures in dry sausage production, *Appl. Environ. Microbiol.*, 40, 472, 1980.
9. Raccach, M. and Baker, R. C., Formation of hydrogen peroxide by meat starter cultures, *J. Food Protect.*, 41, 798, 1978.
10. *Good Manufacturing Practices, Fermented Dry and Semi-Dry Sausage*, American Meat Institute, Washington, D.C., 1982.
11. Coretti, K., Starter cultures in the meat industry, *Fleischwirtschaft*, 3, 386, 1977.
12. Niskanen, A. and Nurmi, E., Effect of starter culture on staphylococcal enterotoxin and thermonuclease production in dry sausage, *Appl. Microbiol.*, 34, 11, 1976.
13. USDA, The Staphylococcal Enterotoxin Problem in Fermented Sausage, Task Force Report, Food Safety and Quality Service, USDA, Washington, D.C., October 1977.
14. National Center for Disease Control, Staphylococcal food poisoning associated with Genoa and hard salami — United States, *Morbidity Mortality Wkly Rep.*, 28, 179, 1979.
15. *Prevention of Microbial and Parasitic Hazards Associated with Processed Foods*, National Academy of Sciences, Washington, D.C., 1975, chap. 7.
16. Bartholomew, D. T. and Blumer, T. N., Inhibition of *Staphylococcus aureus* by lactic acid bacteria in country-style hams, *J. Food Sci.*, 45, 420, 1980.
17. Bacus, J., Reduces nitrosamines, *Food Eng.*, 51, 24, 1979.
18. Gilliland, S. E., Use of lactobacilli to preserve fresh meat, *Proc. Recip. Meat Conf.*, 33, 54, 1980.
19. Sirvio, P. and Nurmi, E., The effect of starter cultures and various additives on the growth of *Salmonella seftenberg* in dry sausage, *Fleischwirtschaft*, 5, 1007, 1977.
20. Christiansen, L. N., Tompkin, R. B., Shaparis, A. B., Johnston, R. W., and Kautter, D. A., Effect of sodium nitrite and nitrate on *C. botulinum* growth and toxin production in summer style sausage, *J. Food Sci.*, 40, 488, 1975.
21. Taylor, S. L. Leatherwood, M., and Lieber, E. R., A survey of histamine levels in sausages, *J. Food Protect.*, 41, 634, 1978.
22. Eitenmiller, R. R., Koehler, P. E., and Reagen, J. O., Tyramine in fermented sausages: factors effecting formation of tyramine and tyrosine decarboxylase, *J. Food Sci.*, 43, 699, 1978.
23. Rice, S., Eitenmiller, R. R., and Koehler, P. E., Histamine and tyramine content of meat products, *J. Milk Food Technol.*, 38, 256, 1975.
24. Nitrite in Meat Curing: Risks and Benefits, Council for Agricultural Science and Technology, Rep. No. 74, CAST, Ames, Iowa, March 6, 1978.
25. Sen, N. P., Tyengar, J. R., Donaldson, B. A., and Poralaks, T., Effect of sodium nitrite concentration on the formation of nitrosopyrrolidine and dimethylnitrosamine in fried bacon, *J. Agric. Food Chem.*, 22, 540, 1974.
26. Tanaka, N., Traisman, E., Lee, M. H., Cassens, R. G., and Foster, E. M., Inhibition of botulinum toxin formation in bacon by acid development, *J. Food Protect.*, 43, 450, 1980.
27. Houston, D., Acid-producing microorganisms in meat products for nitrite dissipation, *Fed. Regist.*, 44, 9372, 1979.
28. Brown, W. L., Starter cultures — new vs. old techniques, in *Meat Ind. Res. Conf.*, American Meat Institute Foundation, Washington, D. C., 1980.
29. Dethmers, A. E., Rock, H., Fuzio, T., and Johnston, R. W., Effect of added sodium nitrite and sodium nitrate on sensory quality and nitrosamine formation in thuringer sausage, *J. Food Sci.*, 40, 491, 1975.
30. USDA, Study to Survey Nitrosamine Levels in Dry Cured Bacon, Hams, and Shoulders, Food Safety Quality Service, USDA, Washington, D. C., February 11, 1980.

31. Leistner, L. and Rodel, W., The significance of water activity for microorganisms in meats, in *Water Relations of Foods*, Duckworth, R. B., Ed., Academic Press, New York, 1975, 309.
32. Acton, J. C. and Dick, R. L., Improved characteristics for dry fermented turkey sausage, *Food Prod. Dev.*, 9, 9194, 1975.
33. Klettner, P. G., Firmness changes during raw sausage aging, *Fleischerei*, October 1980.
34. Acton, J. C., The chemistry of dry sausages, in *Proc. 30th Annu. Recip. Meat Conf.*, American Meat Science Association, Auburn, Alabama, 1977.
35. De Ketelaere, A., Demeyer, D., Vanderkerckhove, P., and Vervaeke, J., Stoichiometry of carbohydrate fermentation during dry sausage ripening, *J. Food Sci.*, 39, 297, 1974.
36. Zaika, L. L., Zell, T. E., Palumbo, S. A., and Smith, J. L., Effect of spices and salt on fermentation of Lebanon bologna-type sausage, *J. Food Sci.*, 43, 186, 1978.
37. Acton, J. C., Dick, R. L., and Norris, E. L., Utilization of various carbohydrates in fermented sausage, *J. Food Sci.*, 42, 174, 1974.
38. Ingolf, F. N. and Skjelkvale, R., Effect of natural spices and oleoresins on *Lactobacillus plantarum* in fermentation of dry sausage, *J. Food Sci.*, 47, 1618, 1982.
39. Zaika, L. L. and Kissinger, J. C., Fermentation enhancement by spices: identification of active component, in *Proc. 42nd Annu. Inst. Food Technol. Meet.*, Las Vegas, 1982.
40. Chaiet, L., Method of Fermenting Meat Products and Composition There For, U.S. Patent 2,945,766, 1960.
41. Salzer, U. J., Broeker, U., Klie, H. F., and Liepe, H. U., Effect of pepper and pepper constituents on the microflora of sausage products, *Fleischwirtschaft*, 57, 2011, 1977.
42. Donnelly, L. S., Ziegler, G. R., and Acton, J. C., Effect of liquid smoke on the growth of lactic acid starter cultures used to manufacture fermented sausage, *J. Food Sci.*, 47, 2074, 1982.
43. Raccach, M., Method and Bacterial Compositions for Fermenting Meats, U.S. Patent 4,303,679, 1981.
44. Demeyer, D., Vandekerckhove, P., Vermeuler, L., and Moermann, R., Compounds determining pH in dry sausage, in *Proc. Eur. Meet. Meat Res. Workers*, Belgrade, Yugoslavia, 1978.

Chapter 6

THE PROPIONIBACTERIA: MILK PRODUCTS

George W. Reinbold

TABLE OF CONTENTS

I. INTRODUCTION

Carbon dioxide and flavor compounds resulting from the growth of certain species of the genus *Propionibacterium* may be either beneficial or deleterious during the ripening of certain popular cheese varieties. Pigmented spots in cheese caused by localized growth of one of the four separate species of dairy-related propionibacteria further emphasize the need for understanding the use of propionibacteria as bacterial starter cultures in cheese. Their importance is not restricted to utilization as cheese starters alone. Many references to their usability and nutritional value in fermented dairy product-based drinks may be found in European literature. Numerous processes have been described for their use in the production of vitamin B_{12} and propionic acid. Some interest has centered on their significance as intestinal tract bacteria and on their possible contribution to silage fermentation. Although not known as a practical application in the U.S., it has been claimed that the addition of a culture of propionibacteria to fat permits prolonged storage by reducing oxidative effects. The variety of possible uses makes the propionibacteria a worthwhile subject for discussion. In truth, there has been comparatively little attention given to them in the dairy and food literature as starter cultures compared with the mesophilic lactic streptococci. In regard to their most important present-day usage in foods, that is to say cheese, contemporary improvements in farm sanitation, milking practices, and refrigeration have increased the necessity of understanding and using carefully selected and nurtured cultures of propionibacteria.

Most of the references in this discussion that normally would have been cited directly have appeared in either one or several of eight previously published literature reviews. These reviews are found in References 1 to 8. To avoid bothersome iteration of the many previously cited specific references in a general discussion of this material they will not be repeated. To date, the use of propionibacteria as bacterial starter cultures for foods in the U.S. has been confined exclusively to their utilization in cheese. Although they have been used in combination with other organisms in Kefir, this use is not commercially practiced in this country. For this reason, this discussion will be restricted to their usage only in cheese.

II. SPECIES INVOLVED

Reorganization of the genus *Propionibacterium*, as published in the eighth edition of *Bergey's Manual of Determinative Bacteriology*,[9] has improved a previously inadequate and confusing system. Eleven species formerly considered as members of the family Lactobacteriaceae, tribe Lactobacilleae (thereby making them unusual next-door neighbors to the genera *Lactobacillus, Microbacterium*, and *Butyribacterium*) have been assigned to the Coryneform group where they (with the *Eubacterium*) now constitute the two genera of the family Propionibacteriaceae. As a result, the dairy-associated propionibacteria at this time share common classification with four species that were formerly members of the genus *Corynebacterium*.

The presently recognized four species of *Propionibacterium* that are dairy-related include *P. freudenreichii* (comprising the subspecies *freudenreichii, globosum*, and *shermanii*), *P. theonii, P. acidi-propionici*, and *P. jensenii*. The habitats of the four remaining species (*P. avidum, P. acnes, P. lymphophilum*, and *P. granulosum*) have been described as blood, pus, abscesses, and wounds. *P. jensenii* is the only dairy-associated species that occasionally shares these latter habitats.

To specify identifying characteristics of the *Propionibacterium*, it is almost obligatory to cite *Bergey's Manual of Determinative Bacteriology*:[9] "Gram-positive, non-spore-forming, non-motile rods. Usually pleomorphic, diphtheroid or club-shaped with one end rounded and the other end tapered or pointed and stained less intensely. Cells of some cultures may be coccoid, elongate, bifid or even branched. Cells usually arranged in singles, pairs or V and Y configurations, short chains or clumps in 'Chinese character' arrangement. Chemoorganotrophs: Metabolize carbohydrates, peptone, pyruvate or lactate. Fermentation products include combinations of propionic and acetic acids and frequently lesser amounts of isovaleric, formic, succinic or lactic acids and carbon dioxide. All species produce acid from glucose. Anaerobic to aerotolerant." This generalized delineation, however, encompasses the four species formerly regarded as corynebacteria and therefore, should be more precisely drawn here to their exclusion. Their inclusion was based primarily on the bases of cell wall and DNA composition, general fermentation products, anaerobic and aerotolerant tendencies, microscopic appearance, and catalase production. Some distinctions separating the two groups (dairy types vs. nondairy types) are

1. A tendency for the DNA of the dairy types to have higher G + C (guanine plus cytosine) content (65.25 to 67.5 vs. 58.25 to 60.00 mol%).
2. Lower average terminal growth pH values in glucose broth for the dairy group (4.43 to 4.90 vs. 4.8 to 5.58).
3. Usual habitats as noted previously.
4. Lack of hydrolysis of gelatin by the dairy group (negative vs. positive).
5. Esculin hydrolysis (4/4 positive vs. 3/4 negative).
6. More strains of the dairy types appear to coagulate milk; this activity is supported by their greater ability to ferment lactose and would be in line with their general habitat.

Isolation of propionibacteria from fermented dairy products, notably cheese, presents no particular problems. Use of a candle-oats jar, relatively oxygen-impermeable pouches, clear-agar covered plates, or comparable techniques with a sodium lactate-containing medium promotes good growth of propionibacteria and will inhibit growth of more nutritionally fastidious organisms. With experience, typical cream to light tan pigmented colonies assuming lenticular and/or "butterfly" shapes can be readily distinguished from the other colony forms. Tentative identification as propionibacteria depends on the ability of the purified cultures to produce acid from lactate, be catalase positive, and characteristically have a clear, growth-free layer on the surface of undisturbed liquid cultures in broth media. Admittedly, this is not a faultless procedure, but with the help of the restrictive environment of a natural cheese it is quite dependably selective.

III. METABOLIC ACTIVITIES

The metabolic activities of the propionibacteria are described in detail in the reviews by Hettinga and Reinbold[1-3] and need not be discussed here in such detail.

In earlier years, the propionibacteria were considered to be essentially the sole procreators of flavor compounds in "eyed" cheeses (excluding those cheeses in which the eyes were formed by *Leuconostoc* or *Streptococcus lactis* subsp. *diacetylactis*). This partially accounts for the great interest in pathways leading to the formation of propionate, acetate, and CO_2 in these cheeses. von Freudenreich and Orla-Jensen are credited with first isolating the organism responsible for the propionic and acetic acid fermentation in Swiss cheese in 1907. In time, it was noted that Swiss-type cheese contained significant amounts of proline. Although these compounds are responsible for a great deal of the characteristic flavor and appearance of these cheeses, the biochemistry of the propionibacteria is very diverse and they form other compounds that may be equally important in developing a fully balanced "Swiss" cheese flavor. The propionibacteria, however, only function properly in cheese through a one-sided commensalism in which the lactic-acid bacteria are the profitless partners. Lactate is usually the driving force in most of their significant fermentations. Lactose may be fermented by most species and strains of propionibacteria other than *P. freudenreichii* subsp. *freudenreichii*, but acidulation is comparatively weak. Glucose, galactose, citrate, pyruvate, and other substrates in milk and cheese may be metabolized and transformed into a number of end products, however, a rich source of lactate must always be present to support their active growth and metabolic processes in cheese.

Biede et al.[10] elucidated their beliefs in this regard in 1973. They fermented cream singly and with combinations of *Lactobacillus bulgaricus, Streptococcus thermophilus,* and *P. shermanii,* roughly following a Swiss cheese-making and ripening schedule of time, temperature, and pH. Through centrifugal removal of the fat, petroleum ether and water extraction, and lyophilization of the water extract, three flavor-bearing fractions were obtained: oil, water-soluble volatiles, and water-soluble nonvolatiles. They believed that the water-soluble nonvolatile fraction contained chiefly amino acids, lactic acid, peptides, and salt. This fraction showed that the typical sweetness of Swiss cheese depends on an interaction of calcium and, possibly, magnesium ions with amino acids and small peptides. These obvious end products of proteolysis were also responsible for burned and bitter flavors. Both the water-soluble volatile and oil-soluble fractions contained volatile and nutty flavors associated with short and medium-chain free fatty acids. In the oil phase, the flavor of the free fatty acids was greatly changed by the presence of lactones, carbonyls, and phenolic compounds. They feel, and most logically, that lipolysis, proteolysis, and the production of acetic and propionic acids must interact to determine the amounts and distribution of the important flavor compounds.

A. Mixed Fermentations

Initial explanations of the role of propionibacteria in cheese flavor production are now known to have been unrealistically simplified. Flavor precursors and compounds other than acetic and propionic acid plus proline are involved. Then, too, the function of the lactic acid-producing bacteria (streptococci and lactobacilli) in the very complex fermentations occurring in cheese was understated. They are indeed required to ferment most of the milk sugar present to produce lactic acid thereby promoting proper curd characteristics, affording control of undesirable fermentations, and inducing slight but necessary protein hydrolysis. We now recognize that they directly contribute to flavor production and greatly control the subsequent growth of the propionibacteria. The symbiotic relationships that exist between *S. thermophilus* and certain lactobacilli (*L. bulgaricus* and *L. helveticus*) will be discussed in other chapters of this book. It should be sufficient to state here that practical cheesemakers have long known that careful selection of functional and compatible strains of lactic acid-producing bacteria is necessary for the consistent production of high quality cheese. Initial proper pH control and lowered O/R potential is necessary to permit the propionibacteria to predominate in subsequent fermentations. Excessive production of CO_2 by heterofermentative activity of some strains of *S. thermophilus* during early curd formation is believed to result in open-textured cheese.

Using commercially sized vats, Biede et al.[11] showed that as the amount of *L. bulgaricus* culture was drastically increased in vat milk, the ensuing cheese contained decreased numbers of propionibacteria, increased free proline, greater proteolysis, and more intense flavors. Diacetyl, propionic acid, and acetic acid levels were not changed significantly by the amount of *L. bulgaricus* added. Cheeses such as Samsoe or Jarlsberg, normally made with *Streptococcus cremoris* and *S. lactis* as well as propionibacteria and which have water added during cooking, have a "Swiss-like" flavor, but invariably a lower propionic to acetic acid ratio resulting in different flavor than in Swiss cheese.

Because it is impossible to prevent the growth of adventitious bacteria in ripening cheeses, their metabolic processes may also influence the growth of propionibacteria. The growth of *Lactobacillus casei*, for example, may increase the growth of propionibacteria, thus positively enhancing flavor and eye development. Microccocci, staphylococci, and enterococci may induce stimulation, inhibition, and even autolysis of propionibacteria. Micrococci, in particular, have been held responsible for enhancement of CO_2 production, thereby giving rise to undesirable increases in CO_2 production during cold storage. Stimulation has been related to formation of important growth factors such as biotin and pantothenic acid or metabolically available substrates such as lactic acid and peptides.

B. Proteolysis

Without a sufficient amount of breakdown in Swiss cheese, desirable characteristic round, smooth, and shiny "eyes" cannot develop. Since propionibacteria do not appreciably affect gross proteolysis on their own, this must be attributed to other agents. The most likely candidates would be the added organisms *L. bulgaricus* or *L. helveticus* and *S. thermophilus* as well as the coagulants used to "set" the milk. The adventitious lactobacilli and micrococci always present in ripening Swiss cheese also contribute to protein hydrolysis. Dipeptidase, aminopeptidase, and proteinase activities in *S. thermophilus* have been reported and would certainly be present in or would be secreted from the lactobacillus cells. In his studies of the formation of proline by propionibacteria, Langsrud has reported the presence of 12 proteinase and 7 peptidase enzymes, separable by electrophoresis, in cellular extracts of a strain of *P. shermanii*.[7]

C. Lipolysis

Rancidity in Swiss-type cheeses attributable primarily to the undesirable growth of clostridia is not uncommon. Microbial "rennets" used in coagulum formation also have been responsible for the production of rancidity. This problem, however, has been overcome and is not known to occur at present. The short-chained, very volatile fatty acids excluding acetic acid are undesirable when present in too-discernible amounts. It is believed though that propionibacteria may release levels of longer-chained fatty acids sufficient to contribute to typical Swiss-type cheese flavor. Propionibacteria are not generally regarded as being actively lipolytic.

D. Propionate, Acetate, and CO_2 Production

Lactate (or pyruvate) is converted to propionate in a novel manner by propionibacteria. This has been concisely expressed by Mantere-Alhonen[8] as occurring by the following reaction: $3CH_3\text{-}CHOH\text{-}COOH \rightarrow 2CH_3\text{-}CH_2\text{-}COOH + CH_3\text{-}COOH + CO_2 + H_2O$. Intermediates in the reaction include pyruvate which is carboxylated to oxalacetate through a biotin-CO_2 complex. Pyruvate is then reduced to malate, fumarate, and succinic acid which with ATP and Coenzyme A (CoA) is converted to succinyl-CoA. The latter compound is then, through the action of methyl-malonyl-CoA isomerase and a B_{12} coenzyme, converted to methyl-malonyl-CoA. This is decarboxylated to yield CO_2 and propionyl-CoA which is split into propionate and CoA. The products propionate, acetate, and CO_2 are in a 2:1:1 ratio.[2,8] The ratio of propionate to acetate varies with different strains. This is influenced by the characteristics of individual strains, acidity of substrates, amount of lactate, and the addition of carbonate and succinate.

E. Proline Production

This amino acid is considered to be a major contributor to the "sweetness" of Swiss-type cheeses and its presence is associated with the growth of propionibacteria in ripening cheese. Three hypotheses have been advanced by Langsrud et al.[12] to account for its production: (1) general proteolysis, (2) hydrolysis of proline-containing peptides by peptidases, and (3) proline biosynthesis. A commercially used strain of propionibacterium was studied to determine the relative importance of these possible pathways. It was concluded that some proline was produced by hydrolysis of casein, but the production rate was slower than by peptidase action on casein hydrolysates. Proline biosynthesis from arginine probably was more important than from glutamic acid, however, biosynthesis was probably less significant than peptide hydrolysis in producing free proline. The concentration of intracellular proline in the propionibacteria was small compared with free proline produced in the media. Release of proline from peptides coincided with release of cellular proteins and enzymes into the medium. It has been shown that propionibacteria can only proteolyze casein slowly so that proline production in cheese is augmented by prior and coincident growth of lactic acid-producing bacterial cultures as well as other extrinsic forms. Other strains of propionibacteria were then tested for proline production under conditions more related to cheesemaking.[13] Using sodium lactate broth, it was shown that differences in the rate of proline production by different species and strains of propionibacteria that were great at 15 days showed little variation after 35 days. Strain differences in proline production at 3 and 21°C were sizable. The rate of production at pH levels and sodium chloride concentrations commonly encountered in Swiss cheese was temporarily retarded but extended incubation negated the effect. Cupric ions at concentrations found in imported Swiss cheese seemingly partly inhibited some of the pathways of proline production. It could be postulated that this partial inhibition of proline production may be beneficial by permitting other flavor processes to proceed more completely.

F. Miscellaneous Compounds

Several investigators have reported the production of diacetyl and acetoin by propionibacteria. Production of diacetyl was greater at 21 than at 32 and 37°C.[3] Diacetyl production in milk cultures was followed by reduction to acetoin and 2,3-butylenediol. Glucose yielded the highest level of diacetyl but pyruvate and citrate could also be used. In mixed culture studies with *S. lactis,* diacetyl production was enhanced by the rapid drop in pH associated with the growth of streptococci.

Other possible contributors to flavor known to be produced by propionibacteria are acetaldehyde, propionaldehyde, ethanol, propanol, dimethyl sulfide, and isovaleric acid.

Although not involved in the production of flavor and eye formation in various cheese varieties, the production of vitamin B_{12} by propionibacteria is worthy of mention. Numerous processes have been described for this purpose using a number of different media usually of natural origin.

IV. FACTORS INFLUENCING STARTER CULTURE PERFORMANCE

A. Introduction

Historically, most cheeses in which propionibacteria played the major role in flavor production and eye formation were made from raw milk. It was early and clearly recognized that the bacteriological quality of the cheese milk was directly related to the quality of the cheese. At the same time, because of many different factors, a specific bacterial flora was wittingly or unwittingly being cultivated in the local milk supplies. Abetting this bacterial development was plant specialization because of small size and locality in single cheese varieties, lack of adequate refrigeration, use of hard-to-sanitize wooden utensils in the kettles or vats, return of whey to the milk-producing farms, sanitary measures, open vats, and other contributory measures. Propionibacteria, because of their association with the intestinal tract of animals including calves, must have become a significant part of that flora. Indeed, within recent years, *P. freudenreichii* and *P. jensenii* have been shown to be present in human saliva.[8] Even today, propionibacteria can be readily isolated from many ripened cheese varieties other than the Swiss types. Small wonder, then, that an adequate number of propionibacteria were usually provided by the milk itself and the fermented whey starters that were then commonly used. As time passed and cheesemaking practices changed (including the use of heat-treated milk) it became increasingly necessary to assure controlled fermentations through the use of starter cultures of propionibacteria.

In the following discussion, propionibacterial starter cultures will be reviewed from contrasting viewpoints, i.e., their optimal production conditions and vat and cheese usage.

B. Species Involved

Although *P. freudenreichii* subsps. *freudenreichii* and *shermanii* have been most frequently used in cheese manufacture, other species have been found to perform satisfactorily. Major emphasis should be placed on capability to produce desirable flavor and sufficient CO_2 commensurate with cheese variety and mode of manufacture. Pigmented *P. theonii*, obviously, is undesirable.

C. Growth Medium and Conditions

Most growth media for propionibacterial starter culture production contain some combination of sodium lactate, nitrogen sources derived from animal or vegetable protein, and yeast extract as a vitamin and accessory growth factor supply. A commonly used medium contains 1% trypticase, 1% yeast extract, 1% sodium lactate, 0.025% KH_2PO_4, and 0.0005% $MnSO_4$ adjusted to a final pH of 7.0. Skimmed milk or reconstituted nonfat dry milk may be used but must be enriched with yeast extract and a simple sugar such as glucose. Addition of a readily available nitrogen source such as trypticase may be additionally advantageous to promote more rapid growth. Because propionibacteria, in comparison with lactic acid-producing bacterial starter organisms, are comparatively slow growers, the growth media must always be sterilized. An incubation period of 3 days at 32°C is normally sufficient for adequate growth. If milk has been selected as the growth medium, care must be taken to arrest the growth before pronounced coagulation occurs. The coagulum formed has been found to distribute poorly in the vat milk and may float. Mantere-Alhonen,[8] after testing various media for growth potential, stated that trypsin-treated whey mixed with silage press juice gave the best results.

Only minimal amounts of culture need be added to vat milk for desired effects, so bulk starter preparation is usually a laboratory rather than plant function. In fact, cells of starter culture concentrated to a paste form function very well for direct vat additions.

General conditions for growth may be summarized as follows: ammonium nitrate may serve as an adequate nitrogen source as propionibacteria possess the necessary peptidases for synthesis of all essential amino acids. Amino acids, therefore, are not necessary for growth but some of those containing sulfur may be stimulatory. As noted, a medium containing lactate and yeast extract will favor growth but too-high lactate concentrations will inhibit growth. Biotin and pantothenic acid are indispensable vitamins and thiamin has a stimulatory effect on some strains of propionibacteria.

Optimal pH for culture growth lies between 6.5 and 7.0. In cheesemaking if curd pH drops below 5.2 to 5.1, growth initiation will be greatly inhibited. Reduced oxygen tension will enhance growth initiation.

Bergey's Manual of Determinative Bacteriology[9] states that most strains grow in 6.5% NaCl. With dairy-related propionibacteria this is patently untrue. To the contrary, most strains used as starter cultures grow slowly (if at all) in sodium chloride solutions above 3.0 to 5.0%. Sensitivity to sodium chloride varies with speed of lactate fermentation; with the more quickly fermenting strains growth is more strongly inhibited by concentrations above 3.0%.

D. Storage Conditions

Propionibacteria retain activity and high viability for up to 8 weeks when stored at 5°C.[5] Storage at 25°C resulted in a higher biochemical activity but the total number of bacteria decreased. Keeping time at this temperature was about 2 weeks. This is considerably better than the short time the lactic acid-producing starter cultures could be kept, even at 5°C. A factor permitting extended low-temperature storage of propionibacteria is the fact that they are not required to produce large amounts of acid on schedule in a few hours. Coagulation of vat milk and early establishment of Swiss-type curd characteristics are not dependent upon the growth of propionibacteria. If, however, curd pH, moisture level, and O/R potential are favorable only a comparatively few viable cells of propionibacteria need be present initially. Slime and/or capsule formation, a frequent by-product of propionibacterial growth, does not appear to afford protection during low-temperature storage.

E. Growth Temperatures

Bergey's Manual of Determinative Bacteriology[9] states the following, "Growth most rapid at 30—37°C and pH near 7. Some strains grow at 25 and 45°C". This statement falls short of full revelation. Optimal incubation temperature is generally 30 to 32°C for the development of cultures to be used in cheesemaking. Warm room temperatures for development of eyes in various cheeses are usually held within a range of 20 to 25.6°C. Substantial growth is possible, however, over a 4-month period at 2.8 to 7.2°C. This ability has been demonstrated for a substantial number of commercially used strains and carries significant economic importance. For example, Hettinga et al.[14] used strains of propionibacteria able to grow at 3.8°C in Swiss cheese manufacture. The resulting 3- and 6-month-old cheeses had a high incidence of splits (6 of 13 and 10 of 13 lots, respectively). Using strains lacking this ability, only 1 of 10 and 2 of 10 lots of cheese split after similar ripening periods. Significantly, CO_2 production in the warm room (21°C) period by strains able to grow at low temperatures was approximately twice that of strains lacking this proclivity. The metabolic characteristics of the low-temperature active strains differ from those unable to grow and produce CO_2 at low temperatures.[15] Enzymes of low-temperature growing strains showed greater capacities for activity at both lower temperatures and lower pH. The so-called split defect of Swiss cheese renders otherwise acceptable product unfit for slicing due to the development of cracks or splits throughout the cheese body. Further, a definite correlation between low-temperature growth and the "pink ring" defect in rindless cheeses has been shown. In this condition, a pink or reddish zone forms 1 or 2 cm under the cheese surface. This is not to say, however, that these defects are caused only by propionibacteria.

In an interesting approach to diminish the incidence of the split defect, Hofherr et al.[16] have developed a mutagenesis-penicillin counterselection to obtain cold-sensitive mutants of strains of *Propionibacterium.* Seven mutants were found that had growth rates similar to that of the parent at 32°C, but much more slowly at 14°C. Essentially the same authors[17] have suggested that strains that grow at low temperatures evolve a fatty acid composition consistent with their ability to grow at lower temperatures. However, both low-temperature and nonlow-temperature growing strains have a limited ability to adjust their fatty acid content with changes in growth temperature.

The ability of propionibacteria to produce capsules and loose slime has been known for many years. It is not surprising, in view of the information obtained from studies of other genera, that incubation temperatures less than optimal for propionibacteria (15 and 21°C) should be conducive to the formation of their capsules and slime.

Growth at temperatures over 32°C does not pertain to cheesemaking processes in which propionibacteria are used. Studies referring to growth at 40 and 45°C state that it is extremely slow.[5]

F. Availability of Nutrients in Materials to be Cultured

In the previous section concerning growth media and conditions for growth, a limited discussion of these factors was presented. Even though the metabolism of the propionibacteria is considered complex, the formulation of suitable media for starter culture growth should be no problem. Media containing yeast extract, hydrolyzed proteins, and lactate salts should suffice, assuming proper concentrations and pH level.

With certain alterations and conditions being met, cheese curd will normally provide all necessary nutrients for luxuriant growth of propionibacteria. Let us examine these requirements. Preformed amino acids, although beneficial, are not essential for growth; however, certain cultures grow with difficulty in their absence. Sulfur-bearing amino acids are growth stimulatory. Glutamic and aspartic acids are readily assimilated as are other amino acids. Milk proteins, after having been subjected to the proteolytic effects of lactic acid starter and coagulant, should obviously satisfy all nitrogen requirements.

Formation of lactic acid during cheesemaking provides a suitable fermentable substrate, although too high a concentration will hinder growth and fermentation. Citrate utilization will be enhanced in the presence of a suitable lactate level.

As noted previously, pantothenic acid and biotin are essential for growth; riboflavin and thiamin are considered stimulatory. Different strains may show varied responses to vitamins and accessory growth factors during early growth. Normal cheese curd, however, should satisfy all requirements in these regards.

Trace element needs seem to be satisfied by milk. Propionibacteria require minimal amounts of iron, magnesium, manganese, and cobalt for enzymatic functions. Phosphate is also needed, but not at amounts beyond what would be present in cheese curd.

Preconditioning of cheese curd is of value for promoting propionibacterial growth.[1] Addition of yeast extract was definitely stimulatory, but the greatest stimulation came from cell-free filtrates of milk cultures of *S. thermophilus* and *Lactobacillus* spp. It should be reasonably safe to assume that comparable filtrates of *S. lactis* and *S. cremoris* would be equally functional. Considering the reduction of the O/R potential, transformation of lactose to lactates, protein hydrolysis, release of metabolically stimulatory substances from extra and intracellular sources, and manipulation of pH (all attributable to careful use of lactic acid-producing bacterial starter cultures), the contribution of supplementary cultures can be appreciated.

G. Treatment of Product Prior to Inoculation

A clear distinction must be drawn between treatment of the starter medium as opposed to the cheese vat milk before inoculation. A starter medium must, of course, be selected that will induce and support luxuriant growth of the desired strain(s) of *Propionibacterium*. As with other genera and species of starter organisms, suitable facilities, environment, and equipment must all be present. All the usual precautions from selection of media ingredients through cooling and holding of the fully developed culture must be taken. Because propionibacteria have slower growth rates than the lactic acid-producing starter bacteria, and are poor competitors in the event of accidental contamination, extra care must be taken. Included in such precautions is the unqualified necessity for absolute sterilization of the growth medium.

In regard to cheese vat milk, other values and considerations are involved. A desirable milk supply has both a low bacterial and somatic cell count. Relative freedom from gas-forming, spore-forming, and psychrotrophic bacteria is preferred. Removal of extraneous and cellular matter can be accomplished by clarification. This leads to a decrease in eye number and an increase in eye size and uniformity by reducing the number of flaws in the curd where the propionibacterial-generated CO_2 could later collect. Freedom from inhibitory compounds is necessary to ensure adequate growth of both the acid- and flavor-producing starters. Depending upon the type of "eyed" cheese to be made, either heat-treated or fully pasteurized milk is used. Both the procedures have the advantage of inactivation of much of the undesirable microbial flora. Because most strains of propionibacteria are relatively thermoduric, these procedures, if not extreme, will permit their survival to some extent. If their initial numbers were not excessive, the added selected propionibacteria should overgrow the "wild" strains. However, if *P. thoenii* is present reddish-brown colonies may be apparent in the cheese body after ripening. Another method used to control cheese milk flora is the hydrogen peroxide-catalase procedure. The literature is not clear in regard to the effect of the use of permissible levels of H_2O_2 on propionibacteria. When adequate numbers of these organisms are added to the milk, this point is of little concern. Fat content of the cheese milk is almost always reduced to achieve a fat-in-dry-matter content of 45 to 48% in the cheese. Cheese containing too much fat does not favor the development of bright, shiny, and smooth eyes.

Incompatibility of propionibacteria with properly used specific strains of lactic acid-producing bacteria has not been clearly demonstrated except where production of the antimicrobial agent nisin was involved. Careful selection and use of lactic starters is very important, however, as the subsequent successful growth of propionibacteria is completely dependent upon their controlled activity in the curd preparation steps. Further, significant lysis of the lactic acid starters by bacteriophage during curd making or pressing would likewise prevent establishment of the proper microenvironment for growth of the propionibacteria during ripening.

H. Inhibitors in Various Substrates

Many of the compounds and conditions that could conceivably retard growth of propionibacteria have already been discussed in this chapter. For a more complete discussion see the work by Langsrud and Reinbold;[4-7] particular emphasis is focused on Reference 5, which contains a separate section on inhibition of propionibacteria.

Certain of these compounds and conditions of special importance in culturing propionibacteria in media or cheese curd should be reviewed in this chapter. Sodium nitrate, a permissible additive to cheese vat milk in many other countries, is inhibitory. Again strain variations exist. At pH 7, a sodium nitrate concentration of 80 g/ℓ should not retard growth of propionibacteria. At pH 5.2, however, a "normal" strain grew at concentrations from 10 to 80 g/ℓ, while a fast lactate-fermenting strain could not grow at concentrations greater than 10 g/ℓ.[5]

Imported Emmenthal cheese, depending upon country of origin, is frequently made in copper kettles. When this is the case, there is a natural tendency for the curd to contain more elevated levels of copper than would occur had it been made in stainless steel vats. High levels of copper (18 mg/kg) are harmful to cheese quality, but flavor and eye formation are normal in cheeses containing 5 mg/kg of copper.[5]

Nisin, an antibiotic produced by *S. lactis,* first believed to be a desirable additive to cheese curd for the inhibition of *Clostridium,* is now known to be inhibitory to both *Lactobacillus helveticus* and propionibacteria. Here, as with so many other compounds and conditions, propionibacterial strain differences may be responsible for inconsistent results in ripened cheeses.

Metabolic by-products from *Pseudomonas, Bacillus,* and *Lactobacillus* species may inhibit the growth of propionibacteria. Even normal metabolites resulting from overly active growth lactic acid-producing starter bacteria may inhibit growth. Concentration and pH may both be responsible for the restraining effect.

An immunoglobulin obtained from Swiss cheese whey was inhibitory to 6 of 8 species of *Propionibacterium* (as formerly classified).[5] The pigmented species (then *P. rubrum* and *P. thoenii* and now *P. thoenii*) were inhibited to a lesser degree than the pigmented species.

I. Bacteriophage

There are no known reports of propionibacteria-bacteriophage relationships of any nature. Attempts to isolate propionibacteria phages from various materials at Iowa State University were always unsuccessful. Underlying conditions responsible for this immunity are not known. This characteristic is shared with species of *Leuconostoc.* It is tempting, therefore, to rationalize that the common ability to form extracellular capsules might be responsible. Unfortunately, not all *Leuconostoc* have this trait, nor for that matter do all *Propionibacterium.*

J. Strain-to-Strain Variations

Great disparities between individual strain characteristics exist. This fact was probably primarily responsible for the addition and deletion of "new" species of these organisms over the years. It is difficult to single out any fermentation and/or biochemical reaction that does not vary between strains. Even such a characteristic as the production of catalase varies among strains of *P. acidipropionici.* Pigmentation of *P. freudenreichii* subspp. may range from white to gray or cream, tan, or pink in surface colonies on horse blood agar. Colonies in deep agar may be white, tan or pink. Colonies of *P. thoenii* in deep agar are white, orange, or red. As a group, they may range from anaerobic to aerotolerant or facultative to aerobic. Their microscopic appearance may be greatly different depending upon growth conditions and, of course, strain.

Slime and capsule formation, which seems to be quite characteristic of most dairy-related propionibacteria, may vary greatly. Under the criteria and experimental conditions of an intensive test, at least one strain of each species and 34 of 82 formed slime.[3] Had conditions been changed it is certain that strain reactions also would have changed.

In a study of heat resistance, some strains were found that could survive exposure to 76°C for 10, but never for 20 sec.[5] Almost half of the strains could survive 72°C for 20 sec.

Differences in fermentative speed, amount of resistance to sodium chloride, flavor-producing ability in cheese ripening, cold-temperature growth and resistance, ratio of acetic to propionic acid production, resistance to antibiotics, and many other factors may exist between strains. From a practical standpoint, it is essential that cheese manufacturers screen available strains for best performance under their plant and production procedure conditions.

REFERENCES

1. Hettinga, D. H. and Reinbold, G. W., The propionic-acid bacteria: a review. I. Growth, *J. Milk Food Technol.*, 35, 295, 1972.
2. Hettinga, D. H. and Reinbold, G. W., The propionic-acid bacteria: a review. II. Metabolism, *J. Milk Food Technol.*, 35, 358, 1972.
3. Hettinga, D. H. and Reinbold, G. W., The propionic-acid bacteria: a review. III. Miscellaneous metabolic activities, *J. Milk Food Technol.*, 35, 436, 1972.
4. Langsrud, T. and Reinbold, G. W., Flavor development and microbiology of Swiss cheese: a review. I. Milk quality and treatments, *J. Milk Food Technol.*, 36, 487, 1973.
5. Langsrud, T. and Reinbold, G. W., Flavor development and microbiology of Swiss cheese: a review. II. Starters, manufacturing processes and procedures, *J. Milk Food Technol.*, 36, 531, 1973.
6. Langsrud, T. and Reinbold, G. W., Flavor development and microbiology of Swiss cheese: a review. III. Ripening and flavor production, *J. Milk Food Technol.*, 36, 593, 1973.
7. Langsrud, T. and Reinbold, G. W., Flavor development and microbiology of Swiss cheese: a review. IV. Defects, *J. Milk Food Technol.*, 37, 26, 1974.
8. Mantere-Alhonen, S., Die Propionibakterien der Molkereiindustrie als Darmkanalmikroben, *Meijeritiet. Aikak.*, 40, 1, 1982.
9. Buchanan, R. E. and Gibbons, N. E., Eds., *Bergey's Manual of Determinative Bacteriology*, 8th ed., Williams & Wilkins, Baltimore, 1974, 633.
10. Biede, S. L., Paulsen, P. V., Hammond, E. G., and Glatz, B. A., The flavor of Swiss cheese, in *Developments in Industrial Microbiology*, Society for Industrial Microbiology, Arlington, Va., 1979, chap. 18.
11. Biede, S. L., Reinbold, G. W., and Hammond, E. G., Influence of *Lactobacillus bulgaricus* on commercial Swiss cheese, *J. Dairy Sci.*, 60, 123, 1977.
12. Langsrud, T., Reinbold, G. W., and Hammond, E. G., Proline production by *Propionibacterium shermanii* P59, *J. Dairy Sci.*, 60, 16, 1977.
13. Langsrud, T., Reinbold, G. W., and Hammond, E. G., Free proline production by strains of propionibacteria, *J. Dairy Sci.*, 61, 303, 1978.
14. Hettinga, D. H., Reinbold, G. W., and Vedamuthu, E. R., Split defect of Swiss cheese. I. Effect of strain of propionibacterium and wrapping material, *J. Milk Food Technol.*, 37, 322, 1974.
15. Hettinga, D. H. and Reinbold, G. W., Split defect of Swiss cheese. II. Effect of low temperatures on the metabolic activity of *Propionibacterium*, *J. Milk Food Technol.*, 38, 31, 1975.
16. Hofherr, L. A., Glatz, B. A., and Hammond, E. G., Mutagenesis of strains of *Propionibacterium* to produce cold-sensitive mutants, Unpublished data, 1983.
17. Hofherr, L. A., Hammond, E. G., Glatz, B. A., and Ross, P. F., Relation of growth temperature to the fatty acid composition of *Propionibacterium* strains, Unpublished data, 1983.

Chapter 7

THE PEDIOCOCCI: MEAT PRODUCTS

James N. Bacus and William L. Brown

TABLE OF CONTENTS

I. SPECIES INVOLVED

The first commercially available starter culture for meat products was identified as *Pediococcus cerevisiae.*[1] Although the predominant microflora of fermented sausages at the time consisted of lactobacilli with some pediococci, streptococci, and micrococci, a specific strain was selected from 32 natural isolates of lactobacilli and 9 strains of pediococci as possessing the most desirable physiological characteristics (Table 1).[2] Early researchers chose *P. cerevisiae* strain FPI "due to its rapid growth and resistance to lyophilization". The lyophilization technique was judged the best method at the time to afford distribution of a viable culture as reliable and inexpensive as possible. As a result, this procedure necessitated the use of a strain that was best able to survive the treatment. Strain FPI possessed characteristics in common with both *P. cerevisiae* and *P. acidilactici,* as described in *Bergey's Manual of Determinative Bacteriology.*[3] In addition, another classification scheme indicated that this strain shared characteristics in common with both *P. acidilactici* and *P. pentosaceus.*[2]

Advances in the technology of storing, handling, and shipping frozen materials allowed the subsequent introduction of a frozen culture concentrate in 1968. A revised classification scheme for the genus *Pediococcus* and related microorganisms also indicated that the actual starter culture strain, originally classified as *P. cerevisiae,* was more correctly classified as *P. acidilactici.*[4]

Currently, various strains of *P. acidilactici* and *P. pentosaceus* are the predominant starter cultures utilized for meat products in the U.S. These two species are among the five listed (Table 2) in the Genus *Pediococcus* (Genus III) within the Family Streptococcaceae (Family II) which is classified in "Part 14. Gram-Positive Cocci" according to *Bergey's Manual of Determinative Bacteriology.*[3] The pediococci are described as Gram-positive cocci occurring in pairs or in tetrads as a result of alternate division along two perpendicular planes. They are nonmotile and do not form endospores. The metabolism is fermentative which is homolactic, producing DL-lactic acid; the L (+) enantiomorph generally predominates. Acid, but no gas is produced from glucose, fructose, and mannose while sorbitol and starch are not fermented. Gelatin is not liquefied and nitrates are not reduced to nitrites. The pediococci have complex nutritional requirements and they are microaerophilic. Most strains are catalase negative although some possess nonheme catalase activity. Natural isolates are found in fermenting plant materials and are rare in milk and dairy products. *P. acidilactici* and *P. pentosaceus* are differentiated from *P. cerevisiae* by their ability to grow at both pH 7.0 and at 35°C. They are also microaerophilic while *P. cerevisiae* prefers anaerobic conditions (Table 3).

II. METABOLIC ACTIVITIES IMPORTANT FOR USE AS A STARTER CULTURE

The initial *Pediococcus* starter culture for meat products (*P. acidilactici*) was employed solely for its rapid lactic acid production that would effectively lower the pH of meat. This culture was developed primarily for "summer-type sausages" where it was assumed that "the predominant organoleptic characteristics arose from the individual formulations in the spices, sugar, salt, and meat components as well as contributions from the varied processing schedules as practiced in the individual plants".[5] The tangy flavor that was desirable and common to this class of sausages resulted from a primary lactic acid fermentation. Any organoleptic contributions due to secondary fermentations or other microbial metabolism appeared to be insignificant.[5] However, indigenous microorganisms in the traditional processes without starter cultures could vary in type and number, with corresponding variations in the type and amount of acid

Table 1
PHYSIOLOGICAL CHARACTERISTICS OF *PEDIOCOCCUS CEREVISIAE* STRAIN FPI[a]

Motility	–
Catalase	
APT broth	–
0.05% Glucose	–
CO_2 from glucose	–
Nitrate reduction	–
Litmus milk	No change
Final pH, glucose broth	3.7
Agar shake	Facultative
Acetylmethylcarbinol production	–
Hydrolysis	
Sodium hippurate	–
Esculin	+
Arginine	+
Gelatin	–
Starch	–
Growth	
10°C	+
30°C	+
45°C	+
50°C	+
5% NaCl, 30°C	+
10% NaCl, 30°C	+

[a] Xylose, arabinose, mannose, fructose, galactose, sucrose, trehalose, glucose, salicin, cellobiose, and α-methyl-D-glucoside were fermented. Glycerol was fermented only under aerobic conditions. Lactose, maltose, raffinose, inulin, mannitol, melibiose, rhamnose, sorbose, melezitose, dulcitol, dextrin, sorbitol, and α-methyl-D-mannoside were not fermented.

From Diebel, R. H., Wilson, G. D., and Niven, C. F., *Appl. Microbiol.*, 9, 239, 1961. With permission.

Table 2
IDENTIFICATION KEY FOR THE GENUS *PEDIOCOCCUS*[3]

I. Growth at pH 5.0, not pH 9.0
   A. No growth at pH 7.0 or at 35°C; prefers anaerobic conditions
      1. *P. cerevisiae*
   B. Growth at pH 7.0 and 35°C; microaerophilic
      1. Growth at 50°C
         2. *P. acidilactici*
      2. No growth at 50°C
         3. *P. pentosaceus*

II. No growth at pH 5.0, growth at pH 9.0: microaerophilic
   A. Halophilic
      4. *P. halophilus*
   B. Not halophilic
      5. *P. urinae-equi*

Table 3
CHARACTERISTICS OF PEDIOCOCCI USED AS MEAT STARTER CULTURES[3]

Characteristic	*P. acidilactici*	*P. pentosaceus*
Spheres	0.6—10 μm	0.8—1.0 μm
Optimum temperature	40°C	35°C
Maximum temperature	52°C	42—45°C
Thermal death point	70°C, 10 min	65°C, 8 min
G + C content DNA	44 mol %	38 mol %
Amino acids required	All, except methionine	All
Acid from galactose	+	+
maltose	−	+
sucrose	±	±
Diacetyl produced	+	+
Original source	Sauerkraut, fermenting mash	Mashes, sauerkraut, pickles, silages

production and other metabolic end products. Thus, the desirable tangy flavor was subject to wide and often uncontrollable variation. The utilization of the *Pediococcus* starter culture provided large numbers of the desirable microorganism in the raw product. The subsequent holding of the product at an optimum growth temperature for the starter (43 to 50°C) achieved a rapid, consistent, and desirable flavor and pH in the sausage. This fermentation temperature was recommended by the culture manufacturer to favor the development of the starter culture over the indigenous microorganisms. Traditionally, much lower fermentation, or "greening" temperatures (16 to 38°C) were used by most fermented sausage manufacturers. These conditions favored the development of the indigenous microflora and resulted in extended fermentation times. At the higher temperatures, growth of any indigenous microorganisms was hindered, and the growth of the *P. acidilactici* was favored so that the tangy flavor and pH characteristics were produced essentially by the starter. Thus, the flavor was closely controlled, the fermentation process accelerated, and any food pathogens inhibited by the rapid acid development.[6]

In addition to its lactic acid-producing capability, the strain of pediococci to be employed as a sausage starter culture had to have a minimum salt tolerance of about 5% (essentially in the range of 5 to 8%). Typical sausage blends contain 2.5 to 4.0% sodium chloride; with an initial moisture content of 50 to 60%, a brine level of 4.0 to 8.0% is achieved. Early researchers observed that this salt tolerance was a prerequisite for a successful meat starter culture. Such strains are selectively developed by using laboratory broth cultures containing gradient amounts of salt.[5]

P. pentosaceus has been promoted in recent years as an effective meat starter culture since it has a lower optimum growth temperature than *P. acidilactici.*[7] Although the *P. acidilactici* has achieved considerable success as the first sausage starter, it was originally developed for relatively high-temperature processes (26 to 50°C), as those more often encountered with semi-dry-type sausages. As a result, these strains generate lactic acid at a much slower rate at lower fermentation temperatures ranging from 15 to 27°C, as is commonly employed for the manufacture of dry-type sausages (Table 4). In addition, the use of higher temperatures for the development of *P. acidilactici* was not always desirable to achieve the best product quality and/or safety. As with *P. acidilactici,* the primary metabolic activity important for the use of *P. pentosaceus* as a meat starter culture is the rapid fermentation of an assimilable carbohydrate to lactic acid, thereby lowering meat pH.

Table 4
COMPARISON OF pH VALUES OF GENOA SAUSAGE UTILIZING EITHER *P. PENTOSACEUS* or *P. ACIDILACTICI* AS A STARTER CULTURE[7]

Starter culture	Incubation temp. (°C)	Time (hr)	pH[a]
P. pentosaceus	26.7	20	5.00
P. acidilactici	26.7	38	5.65
P. pentosaceus	29.4	13	5.00
P. acidilactici	29.4	28	5.40

[a] Initial pH 6.0.

The strains of pediococci have been successful as meat starters since they have been readily adapted to meat systems (i.e., salt tolerance) and efficiently produce large quantities of lactic acid. In addition, they lack the undesirable characteristics that might adversely affect product quality. It has been postulated that the requirements of a good meat starter culture are as follows:[8]

1. Salt tolerant
2. Fast growing in 6% brine (% salt/% moisture × 100)
3. Ability to grow well in the presence of 80 to 100 ppm nitrite
4. Optimum growth at 32.2°C with range from 26.7 to 43°C
5. Homofermentative, producing only lactic acid from dextrose
6. Nonproteolytic
7. Nonlipolytic
8. Not produce off-flavors as by-products of fermentation
9. Not pathogenic
10. Inactivation at 57 to 60°C

The pediococci employed generally meet these requirements.

Although the pediococci are utilized primarily for their fermentative ability in meat systems, apparently they also exert inhibitory activity toward undesirable microorganisms.[9] This inhibition certainly can be due to rapid acid production in fermented meats, but considerable research has demonstrated a repression of food pathogens and other undesirable microorganisms in nonfermented products.[10] The data to date are far from conclusive, but they indicate that the pediococci are producing either hydrogen peroxide, organic acids in addition to lactic, antibiotics, and/or bacteriocins to effect the inhibitory activity. However, some data have shown that specific strains of *P. cerevisiae* (probably *P. acidilactici*) do not accumulate sufficient amounts of hydrogen peroxide to account for the observed inhibition.[11]

III. FUNCTIONS OF PEDIOCOCCI IN FERMENTED MEATS

As discussed in Chapter 5, many sausage products undergo a microbial fermentation as part of their processing cycle. Traditionally, environmental factors during the formulation and subsequent processing favored the development of the lactic microflora that was associated with the meat, ingredients, and processing environment. These lactic acid microorganisms fermented the added sugars to primarily lactic acid with a

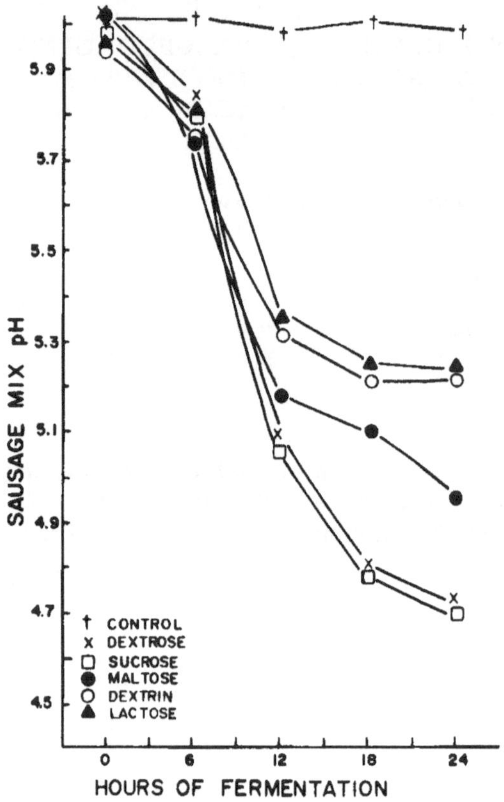

FIGURE 1. Acid production (pH reduction) in fermenting sausage containing 0 or 1% of various carbohydrates. (Reprinted from *J. Food Sci.*, 1977, 42, 174. Copyright © by Institute of Food Technologists. With permission).

corresponding decrease in the pH of the meat. The traditional processes relied upon chance microbial contaminants to inoculate the sausage mix, and the inherent formulation and process controls to ensure a consistent fermentation. As a result, wide product variability was often encountered due to varying types and numbers of the microbial contaminants. In addition, the length of the fermentation cycle would also vary with corresponding alterations in product characteristics.

The use of the *Pediococcus* strains as starter cultures affords a standardized microbial inoculum to each batch regardless of the inherent microflora. The chance contamination still varies, but the inoculum level of the starter culture (recommended at 10^6 to 10^7 microorganisms per gram raw meat) provides the desirable bacteria at such a high level that it generally precludes the development of any chance contaminants and any undesirable effects due to their growth and metabolism.

P. acidilactici initially was employed in the manufacture of semi-dry sausages (summer sausage, thuringer, beef sticks) and some pepperoni where higher fermentation temperatures were more practical. Many processors actually elevated their "greening" temperatures to effect a more rapid fermentation at the starter's optimum growth temperature. As discussed, the primary function was to rapidly lower the pH of the meat without producing by-products that could adversely affect the sausage flavor, texture, and odor. *P. acidilactici* ferments a variety of carbohydrates at varying rates and to various final pH values (Figure 1).[12] Although by definition *P. acidilactici* only produces *slight* acid from sucrose, this characteristic is strain variable and most meat starter cultures readily ferment sucrose since it is encountered typically in sausage products.

The fermentation of dextrose (Table 5)[13] yields primarily lactic acid (Figure 2).[12] The correlation between lactic acid content and pH has been firmly established.[12] Although added carbohydrates are recommended for fermented sausages in order to obtain the necessary pH reduction, the culture will produce small amounts of lactic acid in the absence of added sugars. Apparently, some inherent carbohydrates are present in the meat, and they are also definitely present in most natural spices.

A successful fermentation is not only necessary for the development of the characteristic "tangy" flavor and texture of fermented sausages, but also for the inhibition of spoilage bacteria and food pathogens. The prevalence of pathogenic microorganisms in raw meats and formulation ingredients has been studied extensively.[14] Although highly variable, most common food-borne pathogens can be encountered periodically and should be assumed present. Since many fermented meat products are deliberately held at elevated temperatures during processing to ensure a rapid fermentation, they also provide a desirable environment to accentuate the growth of pathogenic microorganisms. Many of these meat products are not fully cooked, and they usually are eaten by the consumer without further cooking. The ultimate control of product safety must be inherent in the product formulation and process although any risks are reduced by proper sanitation, employee hygiene, and the control of the raw materials. The use of starter cultures provides sufficient microbial numbers to ensure numerical dominance over the natural microflora, including pathogens, and in combination with the proper processing controls guarantees the safety and quality of the final product.

As discussed in Chapter 5, staphylococcal food poisoning caused by defective fermented dry or semi-dry sausage is a potentially significant problem, and several incidents of staphylococcal food poisoning have been traced to fermented sausage where the plant production techniques lacked scientifically based controls designed to ensure the final product was safe for consumption.[15] Specific recommendations, in addition to a continued emphasis on good raw materials and proper sanitation, included controlled acidulation of all fermented sausage with either microbial starter cultures of chemical acidulants. Recently, the industry in the U.S. has published *Good Manufacturing Practices, Fermented Dry and Semi-Dry Sausages* that recommends controlled acidulation via lactic acid-forming microorganisms.[16]

The strains of pediococci as meat starter cultures have proven very effective at inhibiting undesirable microorganisms, including *S. aureus.*[17] Inhibition of staphylococcal growth is enhanced as the ratio of bacteria found in meats to staphylococci increases,[18] and the suppression of staphylococcal growth generally suppresses enterotoxin production (the causative agent in the food poisoning).[19] The large numbers of pediococci inoculated via a starter culture provide a controlled fermentation, accelerate the rate of acid development, and thus indirectly retard the growth of staphylococci. Additionally, certain cultures also exhibit antagonistic effects on staphylococcal growth and/or enterotoxin formation through the production of inhibitory substances and/or competition for essential nutrients.[20]

The pediococci as meat starters are also inhibitory to other pathogenic bacteria such as salmonellae and *Clostridium botulinum*[10] and against undesirable species, including bacilli, Gram-negative enterics, and yeasts. The relative effect generally is dependent upon the species, strain, ratio of lactics to pathogen, the incubation temperature, and the degree and rate of acid production.[21]

Control of the natural fermentation in sausages also appears to prevent histamine accumulation. Histamine has been associated with several outbreaks of food-borne

Table 5
THE ANAEROBIC DISSIMILATION OF GLUCOSE[a]

Reaction	Enzyme	Overall reaction
Glucose + ATP ↓ Glucose-6-phosphate + ADP	Hexokinase	
Glucose-6-phosphate ⇌ Fructose-6-phosphate	Phosphohexose isomerase	Glucose ($C_6H_{12}O_6$)
Fructose-6-phosphate + ATP ⇌ Fructose-1:6-diphosphate + ADP	Phosphohexokinase	+ Glycolytic system of enzymes
Fructose-1:6-diphosphate ⇌ Glyceraldehyde-6-phosphate + dihydroxyacetone phosphate-	Zymohexase (Adolase)	↓
Dihydroxyacetone phosphate ⇌ Glyceraldehyde-3-phosphate	Triosephosphate isomerase	$2CH_3COCOOH$ Pyruvic acid
Glyceraldehyde-3-phosphate + inorganic phosphate + DPN ⇌ 1:3-Diphosphoglyceric acid + $DPN(H_2)$	Phosphoglyceraldehyde dehydrogenase	+ 2H (The hydrogen combines with a hydrogen acceptor, e.g.:
1:3-Diphosphoglyceric acid + ADP ⇌ 3-Phosphoglyceric acid + ATP	Diphosphoglyceric dephosphorylase	$CH_3COCOOH + 2H \rightarrow$ $CH_3CHOHCOOH$) Lactic acid
3-Phosphoglyceric acid ⇌ 2-Phosphoglyceric acid	Triosemutase (Phosphoglyceromutase)	+ energy
2-Phosphoglyceric acid ⇌ 2-Phosphoenolpyruvic acid	Enolase	(The energy is "stored" in phosphate bonds of ATP)
2-Phosphoenolpyruvic acid + ADP ⇌ Pyruvic acid + ATP	Phosphopyruvate dephosphorylase	

[a] Abbreviations: ATP and ADP = adenosine triphosphate and adenosine diphosphate; DPN = diphosphopyridine nucleotide, a co-enzyme involved in hydrogen transfer; and $DPN(H_2)$ = reduced DPN, i.e., DPN that has accepted hydrogen.

illness in the U.S. and numerous episodes abroad. Histamine is formed in foods by the bacterial decarboxylation of histidine and is a normal constitutent of fermented foods, such as cheese, wine, and sauerkraut, and other food products exposed to microbial degradation. Histamine concentrations in fermented meats can be variable and dependent on the length of the aging process.[22-24] Higher histamine levels have been found in dry sausages where a natural fermentation process is employed for an extended aging period. The addition of commercial lactic cultures appears to prevent effectively histamine accumulation through control of the natural fermentation. Apparently, microbial contaminants in the naturally fermented products account for the higher levels of histamine detected.[22] Similar results have demonstrated that use of a starter culture greatly decreases the possibility of development of microflora that possess both tyrosine decarboxylase activity and the proteolytic activity required to produce potentially hazardous tyramine concentrations.[25] Several reports have confirmed

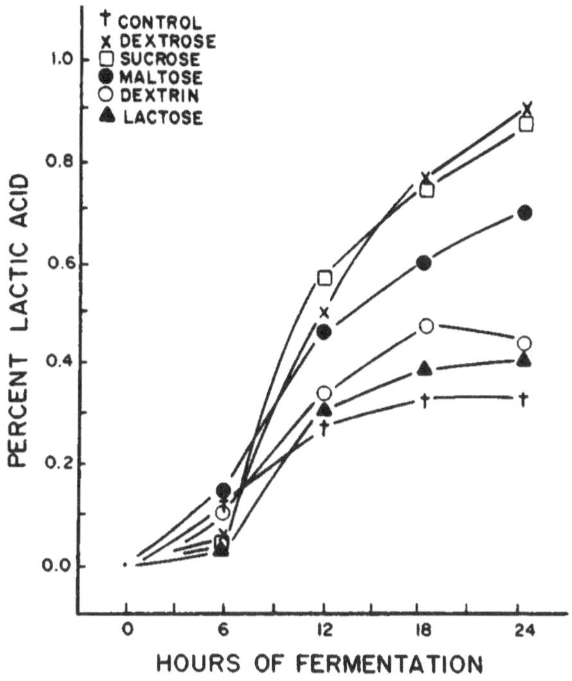

FIGURE 2. Rate of lactic acid accumulation in fermenting sausage containing 0 or 1% of various carbohydrates. (Reprinted from *J. Food Sci.*, 1977, 42, 174. Copyright © by Institute of Food Technologists. With permission.)

the lack of histamine and tyramine production by commercially available *Lactobacillus* and *Pediococcus* meat starter cultures.[26]

Although meat starter cultures have been used primarily in sausage-type products, some research has indicated the beneficial effects in cured meats and typically nonfermented meat products.[10] The pediococci employed in cured meats (i.e., ham) mainly function to inhibit food pathogens and increase the "aged" flavor through lactic acid production.[27] In nonfermented meats such as boiled ham, ground beef, and some poultry products the pediococci appear to exert an antagonistic effect toward undesirable microorganisms, but the critical factor has not as yet been positively identified.[10]

IV. EFFECT OF METABOLITES ON CULTURED PRODUCT

The lactic acid production by the pediococci serves to lower meat pH, typically from 5.6 to 6.2 to 4.7 to 5.2. The lower pH allows the meat to release moisture more uniformly and rapidly since the isoelectric point of meat protein is approximately 5.2 to 5.4.[28] Inadequate fermentation can result in insufficient drying, "case hardening" of the product surface, soft product, a surface "ring", and product "collapsing". It also can result in product spoilage, off-flavors, and a potential health hazard since the initial pH reduction serves to inhibit the growth of most undesirable spoilage microorganisms and food pathogens. This increases product safety and subsequent shelf-life.[21]

In addition to its preservative attributes, the lactic acid denatures meat protein that contributes to the characteristic firm texture of fermented meats. The tangy flavor, also typical of this class of sausages, results from the lactic acid achieved via fermentation and has proven very difficult to duplicate through the use of chemical acidulants.

V. FACTORS INFLUENCING OPTIMUM PERFORMANCE

Meat culture "activity" generally refers to the relative ability of the starter to produce lactic acid in a designated meat system. The rate of pH decline in meat increases with increasing culture "activity". Although the total number of microorganisms definitely influences the fermentation rate, the activity, or acid-producing ability of each cell can be affected by the growth medium, the harvest conditions, the method of preservation prior to subsequent inoculation into the meat, and final meat product formulation and processing parameters.

A. Growth Medium

A commercial culture medium for *P. cerevisiae* (*acidilactici*) has been described as corn steep water, 25 lb; powdered nonfat dry milk, 50 lb; dextrose, 100 lb; yeast autolysate, 13 lb, 5 oz; potassium dihydrogen phosphate, 14 lb, 4 oz; disodium hydrogen phosphate, 9 lb, 9 oz; and water to bring the mixture to 600 gal.[6] The culture medium is inoculated with *P. cerevisiae* (*acidilactici*) in a conventional manner and is incubated for 8 to 10 hr at 32 to 37°C. The bacteria are then separated from the medium (usually by centrifugation), concentrated, mixed with other stabilizing and nutrient materials, and frozen as rapidly as possible. Recommended storage temperature is −20°C or below (optimum −26°C).

Although each commercial culture manufacturer has proprietary media formulations and processing conditions to enhance "activity", it generally has been found that a meat culture must be grown in the presence of sodium chloride (minimum 0.5%) to retain its salt tolerance. Acid-producing capability and salt tolerance are regarded as the two primary attributes of a *Pediococcus* meat starter culture.

Other media components, such as various mineral salts stimulate growth, acid-production, and/or subsequent "activity". In particular, manganese salts have proven very effective to increase the "activity" of both *P. acidilactici* and *P. pentosaceus* strains employed as meat starters.[7]

B. Storage Conditions

Initially, a strain of *P. cerevisiae* (*acidilactici*) was chosen as an appropriate meat starter since it readily survived lyophilization which was the method chosen at the time to preserve and distribute these cultures.[2] However, the "lag phase" required to rehydrate these cultures later proved insufficient, and frozen concentrates were introduced. Frozen cultures are presently the predominant form being utilized in the U.S. Freeze-dried meat cultures still are widely employed throughout the rest of the world, presumably due to problems encountered in distribution of frozen products and the relatively slower fermentation times required.

Culture activity must be sufficiently retained during storage to achieve the necessary performance in the meat system. Various stabilizing agents such as glycerol, nonfat dry milk, monosodium glutamate, malt extract, alkali metal glycerophosphates, glutamic acid, cystine, and/or dextran often are combined with culture concentrate in a freezing medium to afford protection and retain activity.[6] Culture concentrates usually contain 10^9 to 10^{11} cells per gram, and liquid nitrogen or carbon dioxide are employed to freeze the culture as rapidly as possible.

An alternative commercial method to preserve and distribute meat cultures employs a liquid antifreeze agent.[29] A conventional cell paste is diluted with one or more water freezing point depressants that are water soluble, noninjurious to the bacteria, and do not form crystals when cooled to −40°C. Such depressants include polyhydric alcohols, sugars, and other water-soluble inorganic and organic compounds used as cryoprotective agents. The process reduces the water activity of the medium by utilizing a minimum of 40 to 50% of the depressant and provides a nonfrozen concentrate that can be cooled to −40°C to preserve viability and activity. The liquid form enhances the handling characteristics for subsequent use, and the medium inhibits the formation of damaging ice crystals. In addition, the culture concentrate can be warmed during normal distribution without as much loss in viability and activity as encountered with frozen cultures that are thawed.

C. Meat Product Parameters

As discussed in Chapter 5, the meat formulation and process parameters influence both the rate and extent of acid production by the meat starter cultures. Most of the formulation ingredients and environmental parameters affect both lactobacilli and pediococci cultures in a similar manner. To avoid repetition, those parameters will not be discussed.

REFERENCES

1. Deibel, R. H. and Niven, C. F., *Pediococcus cerevisiae:* a starter culture for summer sausage, *Bacteriol. Proc.*, 1957, 14, 1957.
2. Deibel, R. H., Wilson, G. D., and Niven, C. F., Microbiology of meat curing. IV. A lyophilized *Pediococcus cerevisiae* starter culture for fermented sausage, *Appl. Microbiol.*, 9, 239, 1961.
3. Buchanan, R. E. and Gibbons, N. E., Eds., *Bergey's Manual of Determinative Bacteriology*, 8th ed., Williams & Wilkins, Baltimore, 1974.
4. Everson, C. W., Danner, W. E., and Hammes, P. A., Bacterial starter cultures in sausage products, *J. Agric. Food Chem.*, 18, 570, 1970.
5. Deibel, R. H., Niven, C. F., and Wilson, G. D., Microbiology of meat curing. III. Some microbiological and related technological aspects in the manufacture of fermented sausages, *Appl. Microbiol.*, 9, 156, 1961.
6. Rothchild, H. and Olsen, R. H., Process for Making Sausage, U. S. Patent 3,561,977, 1971.
7. Raccach, M., Method and Compositions for Fermenting Meats, U. S. Patent 4,303,679, 1981.
8. Deibel, R. H., Technology of fermented, semi-dried and dried sausages, in *Proc. Meat Ind. Res. Conf.*, American Meat Institute Foundation, Washington, D.C., 1974, 57.
9. Daly, C., Chance, N., Sandine, W. E., and Elliker, P. R., Control of *Staphylococcus aureus* in sausage by starter culture and chemical acidulation, *J. Food Sci.*, 38, 426, 1973.
10. Smith, J. L. and Palumbo, S. A., Microorganisms as food additives, *J. Food Protect.*, 44, 936, 1981.
11. Raccach, M. and Baker, R. C., Formation of hydrogen peroxide by meat starter cultures, *J. Food Protect.*, 41, 798, 1978.
12. Acton, J. C., Dick, R. L., and Norris, E. L., Utilization of various carbohydrates in fermented sausage, *J. Food Sci.*, 42, 174, 1977.
13. Pelczar, M. J. and Reid, R. D., *Microbiology*, McGraw-Hill, New York, 1958, chap. 10.
14. Genigeorgis, C. A., Quality control for fermented meats, *J. Vet. Med. Assoc.*, 169, 1220, 1976.
15. The Staphylococcal Enterotoxin Problem in Fermented Sausage, Task Force Report, Food Safety and Quality Service, U. S. Department of Agriculture, Washington, D.C., October 1977.
16. *Good Manufacturing Practices, Fermented Dry and Semi-Dry Sausages*, American Meat Institute, Washington, D.C., 1982.
17. Raccach, M., Control of *Staphylococcus aureus* in dry sausage by a newly developed meat starter culture and phenolic-type antioxidants, *J. Food Protect.*, 44, 665, 1981.
18. Peterson, A. C., Black, J. J., and Gunderson, M. F., Staphylococci in mixed cultures on growth in artificial medium, *Appl. Microbiol.*, 10, 23, 1962.
19. Niskanen, A. and Nurmi, E., Effect of starter culture on staphylococcal enterotoxin and thermonuclease production in dry sausage, *Appl. Microbiol.*, 34, 11, 1976.
20. Haines, W. C. and Harmon, L. G., Effect of lactic acid bacteria on growth of *Staphylococcus aureus* and production of enterotoxin, *Appl. Microbiol.*, 25, 436, 1973.
21. Bacus, J. N. and Brown, W. L., The use of microbial cultures: meat products, *Food Technol.*, 35, 74, 1981.

22. Taylor, S. L., Leatherwood, M., and Lieber, E. R., A survey of histamine levels in sausage, *J. Food Protect.*, 41, 634, 1978.
23. Cantoni, A., Bianchi, M. A., and Beretta, G., Amino acids, histamine and tyramine variation during ripening of dry sausage (salami),, *Ind. Aliment. (Bucharest)*, 13, 75, 1974.
24. Rice, S., Eitenmiller, R. R., and Koehler, P. E., Histamine and tyramine content of meat products, *J. Milk Food Technol.*, 38, 256, 1975.
25. Eitenmiller, R. R., Koehler, P. E., and Reagen, J. O., Tyramine in fermented sausages: factors affecting formation of tyramine and tyrosine decarboxylase, *J. Food Sci.*, 43, 699, 1978.
26. Rice, S. L. and Koehler, P. E., Tyrosine and histidine decarboxylase activities of *Pediococcus cerevisiae* and *Lactobacillus* species and the production of tyramine in fermented sausages, *J. Milk Food Technol.*, 39, 166, 1976.
27. Bartholomew, D. R. and Blumer, T. N., The use of a commercial *Pediococcus cerevisiae* starter culture in the production of country-style hams, *J. Food Sci.*, 42, 494, 1977.
28. Price, J. F. and Schweigert, B. S., *The Science of Meat and Meat Products*, Food & Nutrition Press, Westport, Conn., 1978, chap. 4.
29. Storrs, A. B., Non-Frozen Concentrated Bacterial Cultures, U. S. Patent 4,226,940, 1980.

Chapter 8

THE LACTOBACILLI, PEDIOCOCCI, AND LEUCONOSTOCS: VEGETABLE PRODUCTS*

H. P. Fleming, R. F. McFeeters, and M. A. Daeschel

TABLE OF CONTENTS

* This is paper no. 9197 of the Journal Series of the North Carolina Agricultural Research Service, Raleigh, N.C. Mention of a trademark or proprietary product does not constitute a guarantee or warranty of the product by the U. S. Department of Agriculture or North Carolina Agricultural Research Service, nor does it imply approval to the exclusion of other products that may be suitable.

I. INTRODUCTION

The use of pure cultures of lactic acid bacteria for fermentation of cucumbers, cabbage, olives, and other produce has been explored for several decades with varying degrees of success. However, pure cultures presently are used only on a limited commercial scale for these commodities. The lack of widespread commercial use is a consequence of several factors, including the following.

1. Cucumbers, olives, and cabbage undergo natural fermentation by lactic acid bacteria if the product is properly handled and held at salt concentrations that have been established for each particular product.
2. Brine from one natural fermentation can be used to inoculate other containers.
3. Heat is the only effective and acceptable means known for ridding vegetables of the natural lactic acid bacteria. Heating is expensive, and it changes the flavor and other characteristics of the product.
4. The fermentation vessels and general handling procedures to date are not compatible with pure culture fermentations.
5. No sufficiently unique strains of lactic acid bacteria have been revealed as yet that have mandated their use as starter cultures.

While each of the above factors may be subject to argument, overall they probably account for the low commercial usage of cultures for vegetables. There are indications, however, that increased use of pure cultures of lactic acid bacteria may occur within the foreseeable future. Changes in brining technology, acceptance of anaerobic fermentation tanks that are more compatible for pure culture usage, and selection and/or modification of lactic acid bacteria with unique and valuable properties loom as possible reasons for use of pure cultures in certain vegetable fermentations.

II. SPECIES INVOLVED

Four species of lactic acid bacteria historically have been associated with the natural fermentation of sauerkraut: *Leuconostoc mesenteroides, Pediococcus cerevisiae, Lactobacillus brevis,* and *L. plantarum*[1] (Figure 1). The latter three species are also associated with fermentation of cucumbers[2] and olives.[3] *L. mesenteroides,* which is very important in the initial stages of fermentation of sauerkraut (approximately 2.25% NaCl), has not been considered as playing an important role in fermentation of cucumbers (5 to 7% NaCl) due to its relatively low salt tolerance. Characteristics of these four species related to vegetable fermentations are summarized in Table 1. These characteristics are largely consistent with *Bergey's Manual of Determinative Bacteriology*[4] (8th edition). The exception are the *Pediococcus* species, which have been classified variously over the past 20 years. *P. cerevisiae* is no longer recognized as a valid species name.[5] Plant pediococci referred to as *P. cerevisiae* by Breed et al.[6] are now called *P. pentosaceus.* For a review of the nomenclature problem, see the review by Garvie.[7] In this chapter, the name *P. pentosaceus* will be used in referring to pediococci present during vegetable fermentations.

III. IMPORTANT METABOLIC PATHWAYS FOR VEGETABLE FERMENTATION CULTURES

A. Fermentation of Sugars

The carbohydrates available for lactic acid fermentation of most fruits and vegetables consist almost exclusively of glucose, fructose, and sucrose. *P. dextrinicus*[8] utilizes

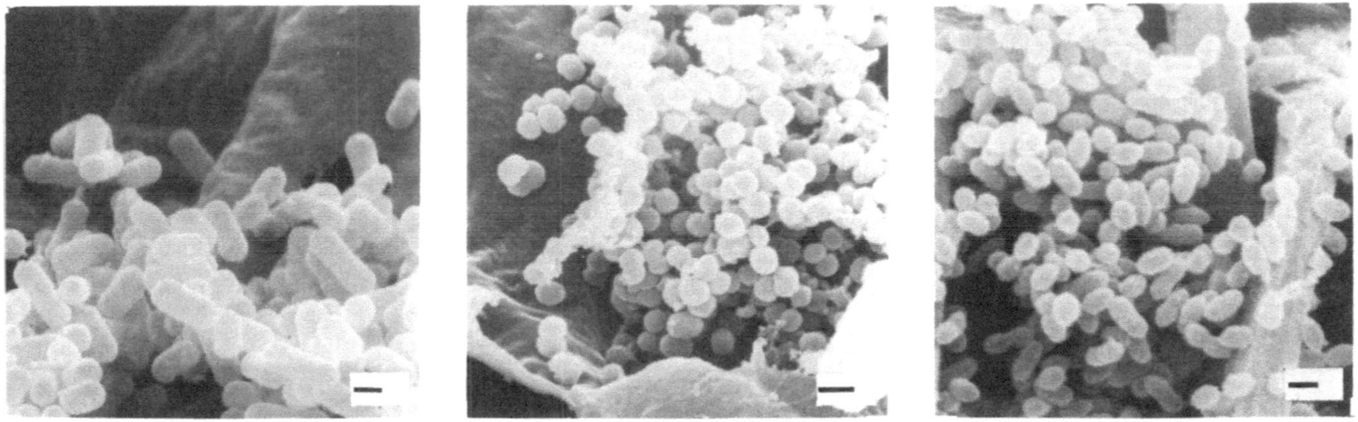

A B C

FIGURE 1. Lactic acid bacteria associated with vegetable fermentations. (A) *Lactobacillus plantarum*, (B) *Pediococcus pentosaceus*, (*P. cerevisiae*), (C) *Leuconostoc mesenteroides*. Bars = 1 μm.

Table 1
RELEVANT CHARACTERISTICS OF LACTIC ACID BACTERIA ASSOCIATED WITH VEGETABLE FERMENTATION

Property	Species: *L. plantarum*	*L. brevis*	*P. pentosaceus*	*L. mesenteroides*	Ref.
Morphology	Short to medium rods, usually singly	Short rods, occurring singly or in short chains	Cocci occurring singly, in pairs, and in tetrads	Cocci or coccobacilli, usually in pairs	4, 105
Optimum temp.	30—35	30	35	20—30	6
Growth at 45°C	No	No	Yes	No	7, 103, 104
Growth in 8% NaCl	Yes	No	Yes	No	4, 7, 103
Lactic acid produced from glucose	DL	DL	DL	D	4, 105
Glucose metabolism	Homofermenter	Heterofermenter	Homofermenter	Heterofermenter	4, 105
Final pH (% acidity)[a] in					
Cabbage juice	3.5 (1.04)	3.9 (1.06)	3.5 (0.90)	3.9 (1.04)	102
Cucumbers	3.2 (0.91)	3.7 (0.54)	3.4 (0.63)	— (0.23)	2, 95
Differentiating biochemical characteristics			Variable		4, 6, 7, 103, 104, 106
Arginine hydrolysis	−	+		−	
Dextran from sucrose	−	−	−	+	
Acid from					
Cellobiose	+	−	+	+	
Sorbitol	+	−	−	−	

[a] Acidity expressed as lactic acid.

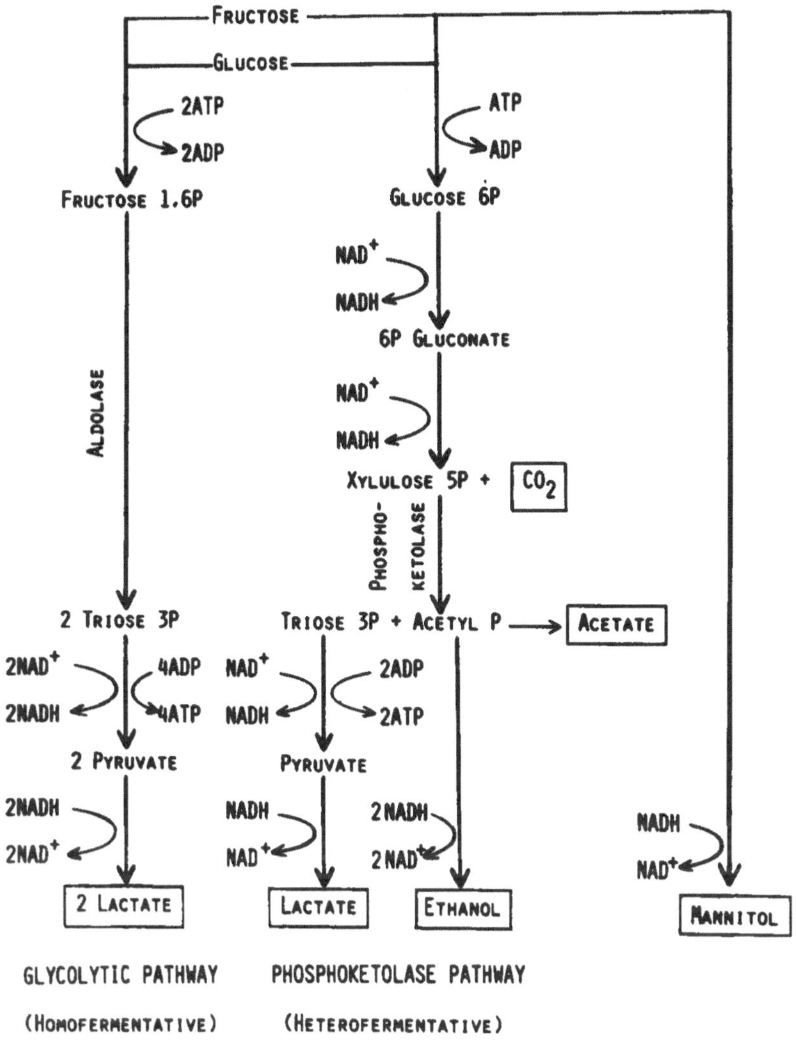

FIGURE 2. Major pathways for sugar fermentation by lactic acid bacteria.

starch, but this bacterium has not been implicated in vegetable fermentations. Free pentoses are not present in vegetables in sufficient amounts to be quantitatively significant in a fermentation. Mannitol may be a significant fermentation substrate in some circumstances. First, it is formed in the heterofermentative phase of sauerkraut fermentation. It then may be a substrate during the homofermentative phase of the fermentation.[1] Second, mannitol is present in substantial amounts in some plants. The most notable case is the common commercial mushroom in which it is the major carbohydrate present.

Lactic acid bacteria involved in vegetable fermentations ferment sugars by two important pathways. Homofermentative lactobacilli, such as *L. plantarum* and pediococci, metabolize hexoses by the glycolytic pathway (Figure 2) to produce primarily lactic acid. A net yield of 2 mol of ATP per mol of hexose fermented is achieved. Electron balance is attained by oxidizing the NADH produced in the transformation of triose phosphate to pyruvate back to NAD by the reduction of pyruvate to lactate. Heterofermentative lactic bacteria (i.e., *L. mesenteroides, L. brevis*) convert glucose to 1 mol each of lactic acid, acetic acid or ethanol, and CO_2 (Figure 2). A net of only 1 mol of ATP is produced per hexose in the 6-phosphogluconate pathway. This pathway also differs from glycolysis in that 3 mol of NADH are formed per hexose. Only 1 mol of NADH is oxidized by reduction of pyruvate. If alternative electron acceptors are present, acetic acid will be a final product, otherwise acetate will be reduced to ethanol to oxidize the other 2 mol of NADH. Fructose can be fermented to the same products as glucose by the heterofermentative pathway in Figure 2, but it can also function as a preferred electron acceptor to oxidize NADH back to NAD^+. The result of this characteristic is that much of the fructose in a heterolactic acid fermentation is reduced to mannitol.

L. plantarum has been classified as a facultative homofermenter[9] because it is capable of producing glucose-6-P-dehydrogenase, 6-P-gluconate dehydrogenase,[9] and phosphoketolase.[10] It has the ability to ferment pentoses using the phosphoketolase enzyme.

There have been only limited quantitative data obtained on the formation of products in vegetable fermentations by either homo- or heterofermentative organisms. Product distributions, fermentation balances, and sugar utilization have been analyzed for *L. plantarum* and 10 heterofermentative species, including *L. mesenteroides* and *L. brevis,* in sterilized green bean juice.[11] The distribution of products was consistent with the homofermentative pathway for *L. plantarum* in that only lactic acid was obtained as a fermentation product; carbon recovery approximated 100%. *L. mesenteroides* and *L. brevis* strains formed typical heterofermentative products. There were strain differences, but measured carbon recoveries were near 100% for *L. brevis* strains and slightly over 100% for *L. mesenteroides.* Carbon recoveries in 7 different vegetables fermented with pH control varied from 74 to 146%,[12] indicating that some substrates or products were unaccounted for in determining the fermentation balances.

As the use of microbial cultures by vegetable fermentation industries increases, it may be important to consider the different characteristics that can be obtained by using either homo- or heterofermentative organisms. The consequences of using a homofermenter are quite simple. Hexoses are efficiently converted to 2 mol of lactic acid, which may be useful if only limited sugar is available. CO_2 is not produced by the glycolytic pathway. Heterofermentative organisms, on the other hand, allow for more complex fermentation possibilities. First, a large amount of CO_2 will be produced from sugars. Second, a more complex product distribution results since acetic acid, ethanol, and mannitol, in addition to lactic acid, are formed. The acid production from a given amount of sugar in a heterofermentation is generally less than that obtained from the same amount of sugar with a homolactic acid fermentation. If we consider, as an example, a case in which equal amounts of glucose and fructose are fermented, and the fructose is quantitatively reduced to mannitol, product formation will be as shown in Equations 1 and 2.

$$2\ \text{Glucose} \rightarrow 2\ \text{Lactic} + 1\ \text{Acetic} + 1\ \text{Ethanol} + 2\ CO_2 \quad (1)$$

$$2\ \text{Fructose} \rightarrow 2\ \text{Mannitol} \quad (2)$$

From a homolactic acid fermentation, 8 mol of acid are formed from 2 mol of glucose and 2 mol of fructose. Equations 1 and 2 show that only 3/8 as much acid is produced on a molar basis by the heterolactic acid fermentation. A result similar to this example has been observed in the fermentation of green beans by *L. cellobiosus.*[13]

B. CO_2 Production

Table 2 lists compounds that can be degraded to yield CO_2 by lactic acid bacteria.

Table 2
REACTIONS FOR CO_2 PRODUCTION FROM ORGANIC AND AMINO ACIDS BY LACTIC ACID BACTERIA

No.	Reaction			Ref.
4	Malate	→ lactate	+ CO_2	15
5	Citrate	→ acetate	+ pyruvate + CO_2	20
6	2 Tartarate	→ lactate	+ acetate + 3 CO_2	107
7	Histidine	→ histamine	+ CO_2	24
8	Tyrosine	→ tyramine	+ CO_2	108
9	Arginine	→ ornithine	+ NH_3 + CO_2	108
10	Glutamic acid	→ α-aminobutyric acid	+ CO_2	24
11	Lysine	→ cadaverine	+ CO_2	24

The degradation of malate is the most important of these reactions due to the widespread ability of lactic acid bacteria to degrade it and the common occurrence of substantial amounts of malate in fruits and vegetables.

Korkes et al.[14] were the first to demonstrate an enzyme in lactic acid bacteria that degraded malic acid with the production of CO_2. They proposed that this "malic enzyme" produced CO_2 by the following reaction:

$$\text{Malic acid} + NAD^+ \rightarrow \text{pyruvic acid} + CO_2 + NADH \quad (3)$$

It was later determined that the reaction found by Korkes et al.[14] in *L. plantarum* was actually a conversion of malic acid to lactic acid and CO_2[15] as shown by Reaction 4 (see Table 2). It was suggested that the enzyme in *L. plantarum* be called L-malate:NAD carboxylyase, or as a common name malolactic enzyme[16,17] to distinguish it from malic enzyme that catalyzes Reaction 3 (above). In Reaction 4, lactic acid is formed instead of pyruvic acid as in Reaction 3.

Caspritz and Radler[18] purified the malolactic enzyme from *L. plantarum* to homogeneity and characterized some of its properties. Surveys of malolactic activity by Caspritz and Radler[18] and malic degradation activity[11,19] in lactic acid bacteria indicate that most strains can produce the enzyme. However, one or more strains that lack the malolactic enzyme have been found in each of the four common vegetable fermentation species, except *L. brevis*.

Though not as widespread as the ability to degrade malic acid, a number of lactic acid bacteria are also capable of CO_2 production from citric acid.[20] In contrast to some earlier reports,[21,22] Keddie[23] found that several cultures of *L. plantarum* isolated from silage degraded citric acid. Citrate is converted to acetate and oxalacetate by citrate lyase. The oxalacetate is then decarboxylated to pyruvate and CO_2. The formation of diacetyl and acetoin from citrate has been the subject of many studies in streptococci, but formation of these compounds has not been investigated in vegetable fermentations.

Amino acid decarboxylation can also be a minor source of CO_2 production by lactic acid bacteria, although in vegetable fermentations it is not established. Rodwell[24] surveyed isolates of rumen lactobacilli and 26 named strains of lactobacilli. Among the named strains, one strain of *L. pentoaceticus* decarboxylated arginine and tyrosine and a strain of *L. bifidus* degraded arginine and glutamic acid. Several of the rumen isolates degraded arginine, histidine, lysine, or ornithine. Radler[17] reported that among 100 isolates of lactic acid bacteria from wines, strains of *L. brevis* could break down arginine, glutamic acid, and isoleucine. One strain of *P. pentosaceus* decarboxylated histidine to histamine.

Table 3
REACTIONS INVOLVING OXYGEN SPECIES IN LACTIC ACID BACTERIA

Reaction	Ref.
$O_2^{-} \cdot + 2\,H^+ + Mn^{2+} \xrightarrow[\text{nonenzymatic superoxide reduction}]{} H_2O_2 + Mn^{2+}$	34, 35
$2\,H_2O_2 \xrightarrow[Mn^{2+}\text{-pseudocatalase}]{} 2\,H_2O + O_2$	38
$\text{Pyruvate} + O_2 + PO_4^{3-} \xrightarrow[\text{pyruvate oxidase}]{} \text{acetyl phosphate} + CO_2 + H_2O_2$	42, 43
$\text{Lactate} + O_2 \xrightarrow[\text{L-lactate oxidase}]{} \text{pyruvate} + H_2O_2$	45
$\text{Lactate} + O_2 \xrightarrow[\text{NAD-independent D-lactate dehydrogenase}]{} \text{pyruvate} + H_2O_2$	45
$\text{NADH} + H^+ + O_2 \xrightarrow[\text{NADH oxidase}]{} \text{NAD} + H_2O_2$	43
$\text{NADH} + H_2O_2 + H^+ \xrightarrow[\text{NADH peroxidase}]{} \text{NAD} + 2\,H_2O$	43

C. Lactic Acid Isomers

Lactic acid bacteria can produce either L(+) or D(−) lactate stereoisomers. Garvie,[25] Stetter and Kandler,[26] and Kunath and Kandler[27] published surveys of the lactic acid isomers produced by this group of bacteria. Among the common species found in vegetable fermentations (Table 1), *L. mesenteroides* produces only the D-isomer, while *L. plantarum, L. brevis,* and *P. pentosaceus* form mixtures of the isomers. It is often said that racemic lactic acid is produced, but in fact equal amounts of the D and L isomer are rarely, if ever, formed. Often the proportion of isomers changes markedly during fermentation.[25,26] The unequal production of isomers is caused by the relative activities of specific D and L lactate dehydrogenases that reduce pyruvate to lactate.[28,29] Hiyama et al.[30] found a second mechanism for formation of mixtures of isomers. *L. sake* produced a lactate racemase that would racemize the L(+) lactic acid initially formed by the organism. However, racemase enzymes have only been found in *L. curvatus* and *L. casei* subsp. *pseudoplantarum* in addition to *L. sake*.[26]

D. Oxygen Reactions

Table 3 shows a list of reactions carried out by lactic acid bacteria that involve oxygen. Lactic acid bacteria have a limited ability to metabolize different forms of oxygen. They are unable to synthesize heme, so they lack cytochromes or heme-containing catalase except in cases in which heme is supplied in the medium.[31-33] *L. plantarum* has also been found to lack superoxide dismutase for dismutation or scavenging of superoxide anion.[34,35] Alternative protective mechanisms against oxygen toxicity have evolved. Scavenging of the superoxide radical is performed by manganous ion in several lactic acid bacteria, including *L. plantarum, P. pentosaceus,* and *L. mesenteroides.*[34,36] To preserve cell viability in late log phase cultures, *L. plantarum* produces a Mn-containing pseudocatalase.[37,38]

Lactic acid bacteria carry out a number of oxidative reactions, usually catalyzed by flavin enzymes.[39,40] *L. delbruckii*[41] and *L. plantarum*[42,43] have a flavin-containing pyruvate oxidase that converts pyruvate to acetyl phosphate and CO_2 with the formation of H_2O_2. The acetyl phosphate can then be utilized for production of ATP. Lactate can be oxidized to pyruvate by two mechanisms that are present in lactobacilli.[44,45] *L.*

curvatus, *L. sake*, *L. acidophilus*, *L. bulgaricus*, and *L. lactis* have a lactate oxidase which reduces O_2 to H_2O_2. *L. plantarum*, *L. casei*, and *L. coryniformis* can use methylene blue or dichlorophenolindophenol as electron acceptors.

Excess electrons can be removed without the production of ATP by a very active NADH oxidase that catalyzes the formation of H_2O_2 and a NADH peroxidase that reduces H_2O_2 to H_2O.[43]

IV. SIGNIFICANCE OF CULTURES IN FERMENTED PRODUCTS

A. General

Unlike liquid foods and beverages, no economically effective methods have been found to completely remove or inactivate the naturally occurring lactic acid bacteria on vegetables prior to fermentation for bulk storage. Heating and other methods have been used on a limited scale, but heating has not been considered practical for large volumes of produce. Other accepted methods such as washing, chlorination, and acidification may reduce numbers, but do not sufficiently attenuate the natural lactic acid bacteria to provide a "pure culture" fermentation. Hence it is important to understand the ecology of natural fermentations in order to appreciate and perhaps develop controlled fermentation methods for vegetables. Considerable research has been done on the lactic acid fermentation of vegetables during this century, including identification of species and characteristics of bacteria involved and environmental factors affecting their predominance during fermentation. Reviews of the natural fermentations of cabbage,[1] cucumbers,[2] and olives,[3,46] are available.

Microbial growth during the natural fermentation of vegetables has been categorized into four sequential stages:[47] *initiation*, which may include growth by many Gram-positive and Gram-negative microorganisms naturally present on the vegetables; *primary fermentation*, which includes growth by lactic acid bacteria with or without growth by fermentative yeasts; *secondary fermentation*, which includes growth by fermentative yeasts after growth by the lactic acid bacteria has been inhibited by low pH, provided that fermentable carbohydrates remain; and *postfermentation*, which occurs after fermentable carbohydrates have been exhausted and is characterized by the absence of microbial growth under anaerobic conditions and by surface growth of oxidative microorganisms only when the brine surface is exposed to the atmosphere. The naturally occurring lactic acid bacteria proliferate during initiation and primary fermentation, depending upon their presence on the raw product and the chemical and environmental conditions under which the brined or salted product is held.

The use of pure cultures of lactic acid bacteria in the fermentation of vegetables and certain fruits in the 20th century has been the subject of numerous studies.[47] Rationale for selection and use of cultures and factors affecting their performance are presented in the remainder of this chapter.

B. Acid Production

Brined or salted vegetables undergo a natural lactic acid fermentation provided the salt concentration does not exceed approximately 8% NaCl and bacterial inhibitors, added intentionally or naturally present in the vegetable, are not present. In natural fermentation of sauerkraut, total acid production and the growth sequence of the four major species of lactic acid bacteria present occur in the approximate order *L. mesenteroides*, *L. brevis*, *P. pentosaceus*, and *L. plantarum*.[1] *L. mesenteroides* characteristically initiates the fermentation and *L. plantarum* characteristically terminates the fermentation. Acid production from these bacteria serves to inhibit growth of acid-sensitive bacteria that greatly outnumber the lactic acid bacteria on the raw plant material.

These species also become inhibited in the order of their sensitivity to acid and/or low pH. Perhaps at least partially for this reason Pederson and Albury[48] found that the fermentation of cucumber brines inoculated with pure cultures of any of the four species were eventually predominated by and terminated by *L. plantarum.*

Lactic acid constitutes most of the acid formed in vegetable fermentations. Acetic acid is also produced, depending on growth by heterofermentative lactic acid bacteria and the oxidation/reduction potential that influences the acetate/ethanol ratio. While lactic acid has a lower pK value, resulting in lower pH values, acetic acid has been shown to be more effective in inhibiting molds associated with softening of brined cucumbers.[49] Acid production with resulting lower pH values is thought to largely account for the eventual predomination of lactic acid bacteria during vegetable fermentations. The extent to which other bacteria grow thus depends upon activity of the lactic acid bacteria.

L. mesenteroides produces only D(−) lactic acid, while most strains of *L. brevis, L. plantarum,* and *P. pentosaceus* associated with vegetable fermentations produce both D(−) and L(+) isomers of lactic acid (Table 1). Both L(+) and D(−) lactic acid are normally metabolized by mammals, but the D(−) isomer is metabolized more slowly.[50] Based primarily on concerns about the ability of infants to metabolize the D(−) isomer, the FAO/WHO Expert Committee on Food Additives[51] recommended that D(−) lactic acid be avoided in infant foods and that consumption of the isomer by adults not exceed 100 mg/kg body weight per day. This recommendation was revised in 1974 to retain the recommendation that D(−) or DL-lactic acid not be used in infant foods, but no limit was set for an acceptable adult intake.[52]

This concern about the consumption of D(−) lactic acid has, however, led to efforts to produce sauerkraut that contains only L(+) lactic acid. Stetter and Stetter[53] found a new species, *L. bavaricus,* isolated from sauerkraut that exclusively forms the L(+) isomer. Use of this bacterium has been tested on an industrial scale to produce up to 50-ton batches yielding sauerkraut with only L(+) lactic acid.[54] A German patent has been granted on the process. *P. dextrinicus* produces exclusively L(+) lactic acid, but it is thought to be found very rarely in nature.[8] The significance of this organism in fermented vegetables is unknown.

C. Sugar Utilization

The rate and extent of sugar utilization during bulk fermentation of vegetables is important for various reasons. Lactic acid produced during fermentation is not subject to further metabolism under anaerobic conditions. If sugars are completely converted to lactic acid during fermentation in bulk containers, it is possible for certain products to be preserved in consumer containers without the requirement for heat processing. The Spanish-style green olive is an example of products that are not heat processed. Fleming et al.[12] concluded that five vegetables (cucumbers, red and green bell peppers, tomatoes, and green beans) fermented by *L. plantarum* with pH control were microbiologically stable under anaerobiosis, provided all fermentable carbohydrates were removed and the pH was 3.8 or below. Failure to completely remove fermentable sugars from carrots and red beets, which had relatively high initial sugar concentrations, resulted in secondary fermentation by yeasts after growth by lactic acid bacteria had ceased, though the pH was maintained at 3.8. Factors that limited growth by *L. plantarum* were not established. If sugar remains after termination of lactic acid fermentation, fermentative yeasts are capable of growing at low pH (<3.8) and under anaerobic conditions. Secondary fermentation by yeasts can result in gaseous spoilage, such as bloater formation in cucumbers.[55] Several factors influence the ability of lactic acid bacteria to completely ferment the sugars of vegetables, including initial sugar concentration, pH, salt concentration, temperature, and buffer capacity of the vegetable. Thus, it is desirable to use cultures of lactic acid bacteria that remove as much fermentable sugar during primary fermentation as possible. *L. plantarum* WSO has been found to be particularly useful for this purpose in fermentation of cucumbers[56] and olives.[57] Even with this culture, however, it was necessary to buffer brined cucumbers with sodium acetate to assure complete sugar utilization.[56] Heterofermentative lactic acid bacteria such as *L. cellobiosus* ferment glucose to acids, but fructose is reduced to mannitol. Chen et al.[58] found that *L. cellobiosus* fermented green beans until all fermentable sugars (primarily glucose and fructose) were removed. However, fructose was reduced to mannitol which could be further fermented by *L. plantarum* WSO above pH 3.6. Below this pH mannitol was not fermented, and the beans were not further fermented by *L. plantarum*. Since mannitol reportedly[59] is not fermentable by yeasts, vegetables fermented to complete hexose removal by heterofermentative lactic acid bacteria may be microbiologically stable even with residual mannitol. This assumption warrants confirmation by further investigation.

The rate of fermentation is important when rapid turnover of the product is desired. Rapid fermentation with brined cucumbers is also desirable from the standpoint of minimizing the time required for purging to remove dissolved CO_2 resulting from fermentation. Rapid fermentation may not be desirable with sauerkraut, however. Higher temperatures may favor rapid fermentation and thus, homo- rather than heterofermentative lactic acid bacteria, which can adversely influence flavor and other quality factors of the product.[60]

D. CO_2 Production

Carbon dioxide production is important in the fermentation of vegetables because of gaseous spoilage problems of the vegetable and stress factors on tank heading devices due to buoyancy pressures. Vegetable tissue and fermenting lactic acid bacteria both contribute to CO_2 production.[61] CO_2 is a major end product in fermentation of hexoses by heterofermentative lactic acid bacteria. However, CO_2 production by homofermentative lactic acid bacteria can occur by decarboxylation of organic and amino acids. Malic and citric acids are present in many vegetables and may be decarboxylated. McFeeters et al.[62,63] found the malolactic reaction to account for most of the CO_2 produced by *L. plantarum* in cucumber juice fermentations.

Perhaps the most notable significance of CO_2 production in brined vegetables occurs with cucumbers. If the critical concentration of CO_2 in the brine is exceeded, the cucumbers will form hollow cavities (bloat) and can represent a serious economic loss (Figure 3d). Bloater damage may be caused by yeasts,[53] coliform bacteria,[64] and homo-[65] and heterofermentative lactic acid bacteria.[66] A mechanism for bloater formation has been proposed[67] that suggests that CO_2 concentration is only one of several factors involved. However, bloater damage can be effectively reduced by purging of CO_2 from the brine.[56,68] Nitrogen is recommended as the purging gas because it will reduce bloater formation and will not cause possible problems due to off-flavors and colors and will not encourage growth of aerobic microorganisms, some of which may produce softening enzymes.[68] Air purging has also been found to prevent bloater formation[68,69] but may result in the problems cited above that do not occur as a consequence of nitrogen purging. Air purging does not appear to adversely influence the growth of lactic acid bacteria at the rates normally used in commercial fermentations, but acid production may be reduced at high rates of air.[49]

Efforts are currently underway to develop strains of *L. plantarum* for use in cucumber fermentations that do not produce CO_2 from malic acid. McFeeters et al.[19] recently observed that bloater damage can be prevented in nonpurged cucumber fermentation if the fermentation is carried out by a homolactic bacterium that lacks the

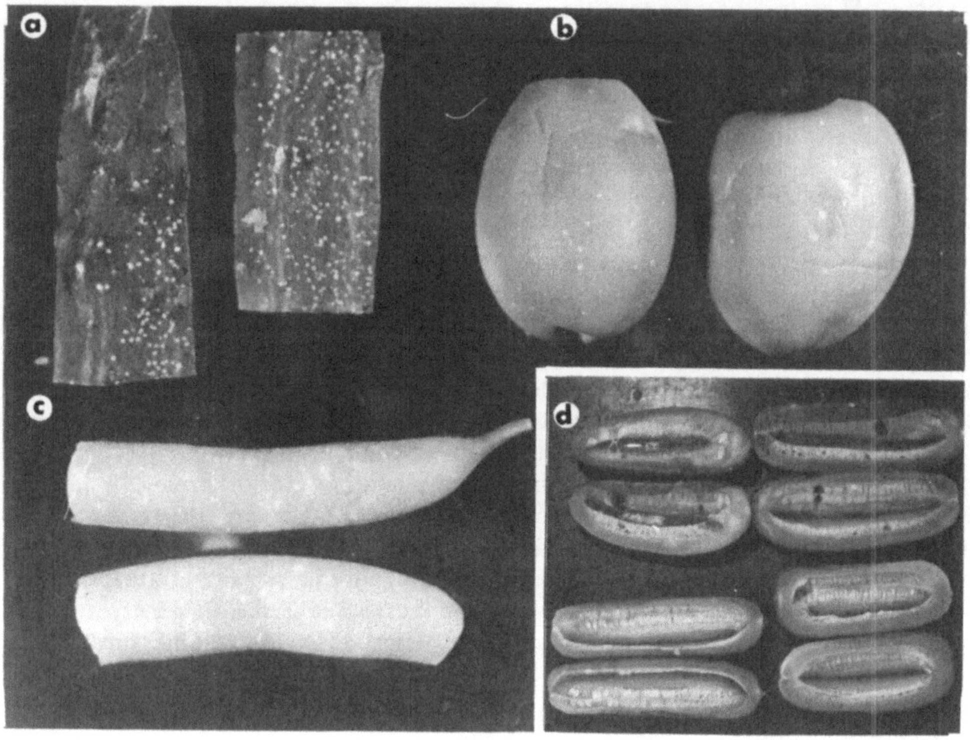

FIGURE 3. Bacterial pustule formation on the surfaces of fermented (a) cucumber, (b) olive, and (c) green bean; (d) gaseous spoilage (bloating) of fermented cucumbers.

ability to degrade malic acid. Predominant growth by such a lactic acid bacterium could obviate the need for purging. Mutants of *L. plantarum* WSO that lack the ability to produce CO_2 from malic acid have been obtained.[70] These and other mutants are being tested and/or further adapted for use in cucumber fermentations.[71]

Gas production, presumably including CO_2, is associated with fisheye spoilage in olives.[72] These lenticular cavities are thought to occur due to spoilage bacteria, but influence due to lactic acid bacteria perhaps should not be excluded.

E. Biogenic Amines

Biogenic amines have been implicated in food poisoning incidents, usually from the consumption of fish. Lactic acid bacteria involved in food fermentations can decarboxylate histidine and certain other amino acids.[73] Mayer et al.[74] investigated histamine formation during sauerkraut production. Commercial sauerkraut samples in the U.S. have been found to contain 0.9 to 13.0 mg/100 g histamine.[75] These concentrations are considered unlikely to cause illness.[75,76] However, it would seem prudent to select cultures that produce minimum amounts of histamine for use in food fermentations.

F. Flavor

Relatively little research has been done on the contribution of lactic acid bacteria to the flavor of fermented vegetables. In the case of fermented Spanish-style green olives, Fleming et al.[77] concluded that the primary contributions of the lactic acid bacteria to flavor were (1) production of a desirable level of acidity and (2) utilization of fermentable sugars to the exclusion of microorganisms that produce end products with undesirable flavor characteristics. They found methyl sulfide to be a major odor component of fermented as well as unfermented olives through headspace vapor analysis. Acetaldehyde and ethanol varied among fermentations by different species of lactic acid bacteria and were found to contribute secondarily to the odor.

Aurand et al.[78] concluded that the flavor of pickles fermented by four different species of lactic acid bacteria in pure culture was due to a blend of volatile components rather than the presence or absence of a single component. The compounds detected by headspace vapor analysis were formaldehyde, acetaldehyde, propionaldehyde, acetone, butyraldehyde, ethyl alcohol, ethyl butyrate, and isovaleraldehyde. The presence and relative concentrations of those compounds varied among species and strains within species. The degree of acidity produced by the cultures evoked the most commentary from the taste panel. *L. plantarum* fermentations were described as strongly acidic and differed from pasteurized controls chiefly in the degree of acid present.

Perhaps the greatest influence of lactic cultures on the flavor of fermented vegetables has been found in sauerkraut. Good-flavored kraut has been associated with the ratio of volatile to nonvolatile acids.[1] Growth by *L. mesenteroides* during the early stages of fermentation is preferred because it forms relatively large amounts of acetic acid. Thus, sauerkraut is preferably fermented at 65°F or below to favor growth of *L. mesenteroides*. At higher temperatures, *L. plantarum* predominates and results in lower ratios of volatile to nonvolatile acid. Also, the pH may be lower and yet the total acid may be lower because of this lower ratio. Such kraut has a sharp acid flavor as opposed to the milder-flavored kraut produced at lower temperatures.

G. Texture

There are indications that high concentrations of lactic acid may result in softening of fermented vegetables. Bell et al.[79] found that the firmness of fresh-pack pickles became progressively softer at higher concentrations of lactic acid during storage. At similar concentrations, lactic acid was more detrimental than acetic acid to firmness. They proposed that the higher binding affinity of lactic acid as compared to acetic acid for Ca^{++} may be responsible for the effects shown. They suggested that removal of Ca^{++} from the pectic substances that bind the cucumber tissues by lactic acid could result in softening. Etchells et al.[57] found Spanish-type green olives fermented by *L. plantarum* to be less firm than those fermented by *L. mesenteroides*. Those fermented by *P. cerevisiae* were intermediate in firmness. Overall, firmness varied inversely with percentage of titratable acidity calculated as lactic acid. These observations indicate that the type of lactic acid fermentation, particularly as it dictates the amount of total lactic acid formed, may influence firmness of the fermented product. Other factors such as salt concentration, Ca^{++} concentration, and storage temperature probably interact with acid concentration to render an overall influence on firmness.

H. Appearance

Lactic acid bacteria influence the appearance of fermented vegetables in the form of brine turbidity, pustules that form on the surface of certain vegetables such as cucumbers and olives (Figures 3a—c), and in gaseous spoilage (Figure 3d). Although cloudy brine is not considered a problem with certain specialty products such as overnight dill pickles, it is not acceptable in most pickle products and in Spanish-type green olives. In genuine dill pickles, the brine must be clarified before it is used as a cover brine in the final product.

V. FACTORS INFLUENCING OPTIMUM PERFORMANCE OF CULTURES

A. Concentrated Cultures

Concentrated cultures of lactic acid bacteria have been commercially available for fermented vegetables for over 10 years. They have been used experimentally in the fermentation of cucumbers[56] and olives,[80] and have been used on a limited commercial scale for the bulk fermentation of brined cucumbers for several years. Both *L. plantarum* and *P. pentosaceus* have been marketed commercially for this purpose. The use of these cultures assures a consistent, rapid fermentation of brined cucumbers if the controlled fermentation procedure of Etchells et al.[56] is followed, the temperature is 26 to 29°C, and the salt concentration does not exceed about 7%.

The cultures are grown and stored by commercial culture firms by general methods used for dairy cultures. These methods are described elsewhere in this book and also have been reviewed by Porubcan and Sellars.[81] The cultures may be stored in liquid nitrogen and are shipped on dry ice. They may be held for several weeks on dry ice or in a −40°C freezer. Such cultures are typically packaged in 70 or 360 g containers with instructions for thawing in about 30°C water. The cells are added at the rate of 1 to 4 billion cells per gallon of brined cucumbers. Before liquid nitrogen storage became popular, cultures were lyophilized. Lyophilization came into disfavor when cell recovery and activity of liquid nitrogen cultures were demonstrated to be superior. Recently, however, attempts have been made to produce improved lyophilized cultures;[81] such cultures are available to the pickled vegetable industry. The lyophilized cultures are shipped without refrigeration and can be stored at about 4°C, or for extended storage at −18°C.

B. Treatment of Product Prior to Inoculation

Cabbage, cucumbers, and green olives are variously handled during preparation for brine fermentation and storage (Figure 4). The growth of the natural flora, including endogenous lactic acid bacteria, is greatly influenced by these treatments.

Most cabbage for sauerkraut is shredded or chopped before tanking, although a small quantity is brined whole or cut into chunks. Thus, nutrients from the cabbage are immediately available and microbial growth probably occurs on the coring and shredding equipment.

Green olives are treated with alkali prior to brining to remove the natural bitterness caused by the phenolic glucoside, oleuropein, in the olives. The olives must then be washed to remove the alkali. The olives may be heat shocked (74°C, 3 min) as proposed by Etchells et al.[57] to enhance the fermentability of certain varieties; normally they are not heated in commercial practice. Sugar may be added to the brine if alkali and washing treatments remove too many of the natural sugars. The brines normally undergo a lactic fermentation by the natural flora, although various abnormal fermentations may result.[3] Brines may be inoculated with an actively fermenting brine, as has been done commercially, or with pure cultures of *L. plantarum*, as has been done experimentally.

Cucumbers are size graded and brined whole. Most commercial fermentations are by naturally occurring lactic acid bacteria. Limited commercial use of lactic cultures has begun during the past decade according to modifications of the controlled fermentation procedure outlined by Etchells et al.[56] This procedure is outlined in Figure 4. Acidification (acetic acid or vinegar) serves to virtually eliminate growth of the natural microbial flora that occurs during the initiation stage of natural fermentations. The brines are purged with nitrogen to remove dissolved CO_2 and thereby prevent bloater formation. After about 24 hr, the brine is buffered with sodium acetate,[56] or the acetic acid is neutralized to form acetate as practiced commercially.[82] The buffering serves to eliminate secondary fermentation by yeasts. With the pH at about 4.7, the brine is inoculated with either *L. plantarum* or *P. pentosaceus* or a combination for a total cell count of 1 to 4 billion cells per gallon of brined cucumbers.

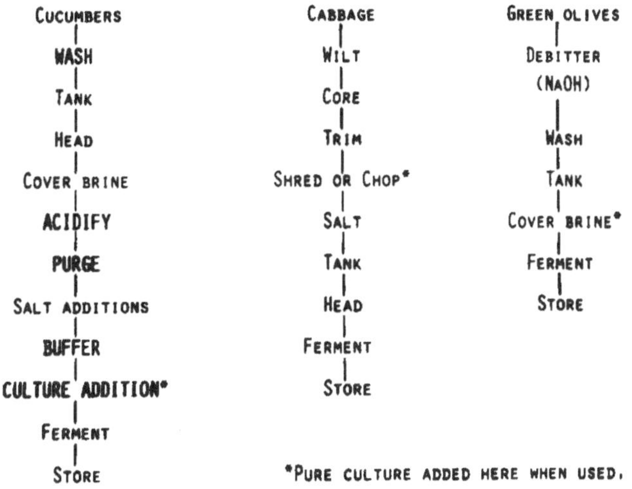

FIGURE 4. Flow charts for the fermentation of produce. For cucumbers, steps that have been added to the natural fermentation procedure are indicated in boldface.[47] For cabbage and olives, steps are those involved in natural fermentation; culture, if used, is added where indicated.

C. Salt Concentration

The use of salt in fermentation serves two primary functions: (1) it helps direct the type and extent of microbial action and (2) it prevents softening of the vegetables. The concentration of salt required to prevent softening determines the minimum that can be used, and varies among products. The concentration for fermentation of cabbage is 2 to 3%; cucumbers, 5 to 8%; and green olives, 4 to 7%. These salt concentrations, in turn, at least partially dictate growth by the four species of lactic acid bacteria. *L. mesenteroides* is active in the primary fermentation of sauerkraut but is not prominent in the fermentation of cucumbers and olives, at least partially because of the higher salt concentrations in the latter products. Currently, there are efforts being made to reduce the concentration of salt in the fermentation of cucumbers to help abate environmental problems and problems related to high salt usage. Addition of calcium acetate[83] or calcium chloride[84] has been shown to result in firmness retention of cucumbers at relatively low concentrations of sodium chloride (1.4 to 4%). It is conceivable that *L. mesenteroides* could be useful in cucumber fermentations at these concentrations of salt. *L. cellobiosus* could also prove to be useful, especially since it was shown to be relatively effective in removing the fermentable sugars of green beans.[13]

D. Temperature

Relatively low fermentation temperatures seem to favor growth of *L. mesenteroides,* while higher temperatures favor growth of *L. plantarum,* which can have important effects on the quality of sauerkraut. Pederson and Albury[60] reported that *L. mesenteroides* predominated an early and extended primary fermentation of sauerkraut at 7.5 and 18°C. At 32 and 37°C, the period of growth by *L. mesenteroides* was shortened and the predominance by *L. plantarum* occurred earlier. They reported that sauerkraut fermented with high salt concentrations or at high temperatures was poor in color, flavor, and texture. Pederson and Albury,[1] based on years of study, concluded that sauerkraut fermented at temperatures of 13 to 18°C will be superior in quality to that fermented at 24°C and above since the heterofermentative lactics exert a great effect at the lower temperatures.

For controlled fermentation of cucumbers, however, Etchells et al.[56] recommended brine temperatures of 26 to 29°C. The homofermentative *L. plantarum* and *P. pentosaceus* cultures ferment rapidly at these temperatures. For controlled fermentation of cucumbers, it has been considered desirable to complete the fermentation as rapidly as possible to reduce purging costs. In the southern and southwestern U.S., brine temperatures during summer months normally exceed 26°C and favor rapid growth of these starter cultures. In cooler regions of the U.S. such as the northwest, and in late fall brine temperatures may be 21°C or lower. For this reason, efforts have been made to obtain cultures capable of growing rapidly at these temperatures. A strain of *P. pentosaceus* has reportedly been developed that is capable of growing more rapidly at the lower temperatures.[85] Since *L. plantarum* is more acid tolerant, however, it is likely to terminate the fermentation. Indeed, this has been our experience.[71]

E. Availability of Nutrients

Fermentable carbohydrates in cabbage and olives include sucrose, fructose, and glucose. Glucose and fructose constitute essentially all of the fermentable sugars in cucumbers and olives. All other essential nutrients for growth of lactic acid bacteria are present in the raw produce, as evidenced by the occurrence of a vigorous natural lactic fermentation in the brined products. Availability of these nutrients for growth of lactic acid bacteria depends upon treatment of the produce before salting or brining. Nutrients are immediately available in cabbage that is shredded or chopped, whereas they must diffuse from cucumbers and olives into the brine before they are available for fermentation in the brine surrounding the produce.

It has been hypothesized that the fermentation of cucumbers occurs solely in the brine.[66] It has been shown, however, that lactic acid bacteria enter and proliferate within brined cucumbers.[86] Entrance into and growth within cucumbers was shown to be influenced by the gas composition of the cucumber before brining. Thus, fermentation evidently occurs within cucumbers and the brine surrounding the cucumbers. Later evidence has indicated that the relative amount of sugar that diffuses into the brine before inoculation influences the relative amount of fermentation that occurs in the cucumbers and in the brine.[87]

While green olives contain 2 to 3% fermentable sugars[57] (which is more than sufficient for adequate acid production), alkali and washing treatments can greatly reduce the amount of sugars available for fermentation. It is assumed that other nutrients would also be removed or altered by these treatments. The degree of alkali penetration can be an important factor in this regard. California processors customarily allow alkali to penetrate to the pit, thereby more completely removing bitterness from the olives. Fermentable sugar is removed to such an extent that glucose is added to assure adequate acid production. In Spain, briners typically allow alkali to penetrate only about three fourths of the flesh distance to the pit. This leaves a slight bitterness that is preferred in these olives, as well as adequate sugar for acid production.

F. Natural Inhibitors

Unidentified substances in cabbage have been reported to inhibit certain undesirable Gram-negative bacteria present on cabbage.[1] They presumably do not inhibit desirable lactic acid bacteria.

Etchells et al.[57] found that heat shocking of lye-treated, green Manzanillo olives greatly increased their brine fermentation by pure cultures of lactic acid bacteria. When the olives were neither lye-treated nor heated, added lactic cultures caused essentially no fermentation; only yeasts grew. They suggested that the heat treatment destroyed a naturally occurring inhibitor of lactic acid bacteria, and that the inhibitor may account for the occurrence of stuck fermentations which are characterized by the absence of lactic acid bacteria and the presence of yeasts in the brine. Subsequent studies revealed that hydrolysis products of oleuropein, including its aglycone and elenolic acid, are inhibitory to lactic acid bacteria but not to yeasts.[88-90] Fleming et al.[90] found that oleuropein was not greatly inhibitory to lactic acid bacteria, contrary to earlier suggestions by Vaughn[3] and Juven et al.[91] In fact, lactic acid bacteria have been found to utilize oleuropein.[92] Fleming et al.[90] suggested that oleuropein is degraded to its aglycone when unheated olives are brined, perhaps by the natural β-glucosidase that Cruess and Alsberg[93] reported, and by further degradation to elenoic acid. Juven and Henis[94] found that oleuropein became more inhibitory when it was treated with β-glucosidase. Perhaps heat inactivates β-glucosidase in the olives, preventing breakdown of oleuropein to yield the inhibitory aglycone when the olives are brined. Apparently not all varieties of olives are inhibitory to lactic acid bacteria. Balatsouras et al.[80] found no evidence that the Conservolea variety of olives popular in Greece contains compounds in concentrations inhibitory to lactic acid bacteria.

Antagonism among species of lactic acid bacteria has been indicated. Etchells et al.[95] found that *P. pentosaceus* delayed the onset of growth by *L. plantarum* in fermenting cucumbers. Fleming et al.[96] found that certain strains of *P. pentosaceus* were inhibitory to other strains of this species, to other species of lactic acid bacteria, and to certain other Gram-positive bacteria.

G. Bacteriophage

We are not aware of any problems that occur due to bacteriophage in *L. plantarum* or *P. pentosaceus* cultures that are used for pure culture inoculations of vegetables. The presence of bacteriophage in *L. plantarum* and other lactobacilli has been reported,[97] but its significance in cultures for vegetable fermentations is yet to be determined.

VI. CONCLUSIONS

Much research has been done over the past 50 years to characterize the lactic fermentation of vegetables. This work has yet to be sufficiently integrated to encourage wide-scale commercial use of controlled fermentation methods. In this regard, the vegetable fermentation industry contrasts with other food and beverage fermentation industries such as the dairy, brewery, and wine industries. Part of the lag in development and commercial acceptance of such methods is caused by the complexities of fermenting particulate material. The vegetable fermentation industries have been able to produce acceptable products using salt and the mechanisms of nature for fermentation control. A major obstacle for the use of pure cultures and controlled fermentation methods has been the fermentation vessel, especially for brined cucumbers.

Recent efforts to develop anaerobic tanks for fermentation and storage of brined cucumbers[98] are expected to make controlled fermentation an attractive option to picklers. It is conceivable that development of strains of lactic acid bacteria with desirable and unique properties will further encourage acceptance of pure cultures. Development of cultures for use in vegetable fermentations that do not produce CO_2 from malic acid, that exclusively produce L(+) lactic acid, and that have other desirable traits are examples of recent progress that hint of possible breakthroughs for the industry in the near future. Rapid progress is being made in understanding the genetic systems of the lactic acid bacteria that are used in the food industry. Much is already known about the lactic streptococci used in the dairy industry. *L. plantarum* and *P. pentosaceus* are only now being studied with respect to their genetic information transfer systems. Conjugal plasmid transfer has been demonstrated between streptococci and pediococci,[99] and recently, plasmids have been observed in *L. plantarum*.[100,101] Rapid progress is likely in understanding how these important vegetable fermentation bacteria can be engineered to have desirable traits for use as starter cultures.

Thus, development of improved controlled fermentation methods for vegetables seems possible, but commercial acceptance will depend on development of superior cultures and suitable tanking and handling procedures to make such methods economically attractive.

ACKNOWLEDGMENTS

This investigation was supported in part by a research grant from Pickle Packers International, Inc., St. Charles, Illinois.

REFERENCES

1. Pederson, C. S. and Albury, M. N., The sauerkraut fermentation, *N. Y. State Agric. Exp. Stn. Bull.*, 824, 1969.
2. Etchells, J. L., Fleming, H. P., and Bell, T. A., Factors influencing the growth of lactic acid bacteria during the fermentation of brined cucumbers, in *Lactic Acid Bacteria in Beverages and Food*, Carr, J. G., Cutting, C. V. and Whiting, G. C., Eds., Academic Press, New York, 1975, 281.
3. Vaughn, R. H., Lactic acid fermentation of cucumbers, sauerkraut and olives, in *Industrial Fermentations*, Vol. 2, Underkofler, L. A. and Hickey, R. J., Eds., Chemical Publishing, New York, 1954, 417.
4. Buchanan, R. E. and Gibbons, N. E., Eds., *Bergey's Manual of Determinative Bacteriology*, 8th ed., Williams & Wilkins, Baltimore, 1974.
5. Skerman, V. B. D., McGowan, V., and Sneath, P. H. A., Eds., Approved lists of bacterial names, *Int. J. Syst. Bacteriol.*, 30, 225, 1980.
6. Breed, R. S., Murray, E. G. D., and Smith, N. R., Eds., *Bergey's Manual of Determinative Bacteriology*, 7th ed., Williams & Wilkins, Baltimore, 1957, 529.
7. Garvie, E. I., Nomenclatural problems of the pediococci, *Int. J. Syst. Bacteriol.*, 24, 301, 1974.
8. Back, W., Elevation of *Pedicoccus cerevisiae* subsp. *dextrinicus* Coster and White to species status *Pediococcus dextrinicus* (Coster and White) comb. nov., *Int. J. Syst. Bacteriol.*, 28, 523, 1978.
9. Buyze, G., Van Den Hamer, J. A., and De Haan, P. G., Correlation between hexose-monophosphate shunt, glycolytic system, and fermentation-type in lactobacilli, *Antonie van Leeuwenhoek; J. Microbiol. Serol.*, 23, 345, 1957.
10. Heath, E. C., Hurwitz, J., Horecker, B. L., and Ginsburg, A., Pentose fermentation by *Lactobacillus plantarum*. I. Cleavage of xylulose-5-phosphate by phosphoketolase, *J. Biol. Chem.*, 231, 1009, 1958.
11. Chen, K. H., McFeeters, R. F., and Fleming, H. P., Fermentation characteristics of heterolactic acid bacteria in green bean juice, *J. Food Sci.*, 48, 962, 1983.
12. Fleming, H. P., McFeeters, R. F., Thompson, R. L., and Sanders, D. C., Storage stability of vegetables fermented with pH control, *J. Food Sci.*, 48, 975, 1983.
13. Chen, K. H., McFeeters, R. F., and Fleming, H. P., Complete heterolactic acid fermentation of green beans by *Lactobacillus cellobiosus*, *J. Food Sci.*, 48, 967, 1983.
14. Korkes, S., del Campillo, A., and Ochoa, S., Biosynthesis of dicarboxylic acids by carbon dioxide fixation. IV. Isolation and properties of adaptive "malic" enzyme from *Lactobacillus arabinosus*, *J. Biol. Chem.*, 187, 891, 1950.
15. Schultz, M. and Radler, F., The "malic enzyme" from *Lactobacillus plantarum* and *Leuconostoc mesenteroides*, *Arch. Mikrobiol.*, 91, 183, 1973.
16. Kunkee, R. E., A second enzymatic activity for decomposition of malic acid by malo-lactic bacteria, in *Lactic Acid Bacteria in Beverages and Food*, Carr, J. G., Cutting, C. V., and Whiting, G. C., Eds., Academic Press, New York, 1975, 29.
17. Radler, F., The metabolism of organic acids by lactic acid bacteria, in *Lactic Acid Bacteria in Beverages and Food*, Carr, J. G., Cutting, C. V., and Whiting, G. C., Eds., Academic Press, New York, 1975, 17.
18. Caspritz, G. and Radler, F., Malolactic enzyme of *Lactobacillus plantarum*. Purification, properties, and distribution among bacteria, *J. Biol. Chem.*, 258, 4907, 1983.

19. McFeeters, R. F., Fleming, H. P., Fleming, H. P. and Daeschel, M. A., Malic acid degradation and brined cucumber bloating, *J. Food. Sci.*, 49, 999, 1984.
20. Du Plessis, L. de W., The microbiology of South African winemaking. VII. Degradation of citric acid and L-malic acid by lactic acid bacteria from dry wines, *S. Afr. J. Agric. Sci.*, 7, 31, 1964.
21. Gibson, T. and Adb-el-Malek, Y., The formation of carbon dioxide by lactic acid bacteria and *Bacillus licheniformis* and a cultural method of detecting the process, *J. Dairy Res.*, 14, 35, 1945.
22. de Man, J. C., Enige eigenschappen van staafvormige melkzuurbacterien geisoleerd vit nederlandse kaas, *Ned. Melk Zuiveltijdschr.*, 10, 240, 1956.
23. Keddie, R. M., The properties and classification of lactobacilli isolated from grass and silage, *J. Appl. Bacteriol.*, 22, 403, 1959.
24. Rodwell, A. W., The occurrence and distribution of amino acid decarboxylases within the genus *Lactobacillus*, *J. Gen. Microbiol.*, 8, 224, 1953.
25. Garvie, E. L., The production of L (+) and D (−) lactic acid in cultures of some lactic acid bacteria, with a special study of *Lactobacillus acidophilus* NCDO 2, *J. Dairy Res.*, 34, 31, 1967.
26. Stetter, K. O. and Kandler, O., Formation of DL-lactic acid by lactobacilli and characterization of a lactic acid racemase from several streptobacteria, *Arch Mikrobiol.*, 94, 221, 1973.
27. Kunath, V. P. and Kandler, O., Der gehalt an L (+)- und D (−)-milchsaure in joghurtprodukten, *Milchwissenschaft*, 35, 470, 1980.
28. Dennis, D. and Kaplan, N. O., D- and L-lactic acid dehydrogenases in *Lactobacillus plantarum*, *J. Biol. Chem.*, 235, 810, 1960.
29. Gordon, G. L. and Doelle, H. W., Production of racemic lactic acid in *Pediococcus cerevisiae* cultures by two lactate dehydrogenases, *J. Bacteriol.*, 121, 600, 1975.
30. Hiyama, T., Fukui, S., and Kitahara, K., Purification and properties of lactate racemase from *Lactobacillus sake*, *J. Biochem.*, 64, 99, 1968.
31. Whittenbury, R., Hydrogen peroxide formation and catalase activity in the lactic acid bacteria, *J. Gen. Microbiol.*, 35, 13, 1964.
32. Bryan-Jones, D. G. and Whittenbury, R., Haematin-dependent oxidative phosphorylation in *Streptococcus faecalis*, *J. Gen. Microbiol.*, 58, 247, 1969.
33. Sijpesteijn, A. K., Induction of cytochrome formation and stimulation of oxidative dissimilation by hemin in *Streptococcus lactis* and *Leuconostoc mesenteroides*, *Antonie van Leeuwenhoek; J. Microbiol. Serol.*, 36, 335, 1970.
34. Gotz, F., Elstner, E. F., Sedewitz, B., and Lengfelder, E., Oxygen utilization by *Lactobacillus plantarum*. II. Superoxide and superoxide dismutation, *Arch. Mikrobiol.*, 125, 215, 1980.
35. Archibald, F. S. and Fridovich, I., Manganese and defenses against oxygen toxicity in *Lactobacillus plantarum*, *J. Bacteriol.*, 145, 442, 1981.
36. Archibald, F. S. and Fridovich, I., Manganese, superoxide dismutase, and oxygen tolerance in some lactic acid bacteria, *J. Bacteriol.*, 146, 928, 1981.
37. Kono, Y. and Fridovich, I., Functional significance of manganese catalase in *Lactobacillus plantarum*, *J. Bacteriol.*, 155, 742, 1983.
38. Kono, Y. and Fridovich, I., Isolation and characterization of the pseudocatalase of *Lactobacillus plantarum*. A new manganese-containing enzyme, *J. Biol. Chem.*, 258, 6015, 1983.
39. Strittmatter, C. F., Electron transport to oxygen in lactobacilli, *J. Biol. Chem.*, 234, 2789, 1959.
40. Strittmatter, C. F., Flavin-linked oxidative enzymes of *Lactobacillus casei*, *J. Biol. Chem.*, 234, 2794, 1959.
41. Hagar, L. P., Geller, D. M., and Lipmann, F., Flavoprotein-catalyzed pyruvate oxidation in *Lactobacillus delbrueckii*, *Fed. Proc. Fed. Am. Soc. Exp. Biol.*, 13, 734, 1954.
42. Sedewitz, B., Schleifer, K. H., and Gotz, F., Purification and properties of the pyruvate oxidase from *Lactobacillus plantarum*, in *Lactic Acid in Foods*, Proc. Neth. Soc. Microbiol. Meet., Wageningen, September 1983.
43. Gotz, F., Sedewitz, B., and Elstner, E. F., Oxygen utilization by *Lactobacillus plantarum*. I. Oxygen consuming reactions, *Arch Mikrobiol.*, 125, 209, 1980.
44. Hochst, M., Untersuchungen zur Laktatoxidation bei Lactobazillen, Dissertation, Universitat Munchen, Munich, 1979.
45. Kandler, O., Carbohydrate metabolism in lactic acid bacteria, *Antonie van Leeuwenhoek; J. Microbiol. Serol.*, 49, 209, 1983.
46. Vaughn, R. H., Lactic acid fermentation of olives with special reference to California conditions, in *Lactic Acid Bacteria in Beverages and Food*, Carr, J. G., Cutting, C. V., and Whiting, G. C., Eds., Academic Press, New York, 1975, 307.
47. Fleming, H. P., Fermented vegetables, in *Economic Microbiology, Fermented Foods*, Vol. 7, Rose, A. H., Ed., Academic Press, New York, 1982, 227.
48. Pederson, C. S. and Albury, M. N., The effect of pure culture inoculation on fermentation of cucumbers, *Food Technol.*, 15, 351, 1961.

49. Potts, E. A. and Fleming, H. P., Prevention of mold-induced softening in air-purged, brined cucumbers by acidification, *J. Food Sci.*, 47, 1723, 1982.
50. Giesecke, D., Fabritius, A., and Wallenberg, P. V., A quantitative study on the metabolism of D (−) lactic acid in the rat and the rabbit, *Comp. Biochem. Physiol.*, 69B, 85, 1981.
51. Specifications for the Identity and Purity of Food Additives and their Toxicological Evaluation: Some Antimicrobials, Antioxidants, Emulsifiers, Stabilizers, Flour-Treatment Agents, Acids, and Bases, 9th Rep. Jt. FAO/WHO Expert Comm. Food Addit., FAO/WHO, Rome, 1966.
52. Anon., Toxicological evaluation of some food additives including anticaking agents, antimicrobials, antioxidants, emulsifiers, and thickening agents, *WHO Food Addit. Ser.*, 5, 1974.
53. Stetter, H. and Stetter, K. O., *Lactobacillus bavaricus* sp. nov., a new species of the subgenus *Streptobacterium, Zentralbl. Bakteriol. Parasitenkd. Infektionskr. Hyg. Abt. 1: Orig. Reihe C;* 1, 70, 1980.
54. Eden-Waren GmbH, *Lactobacillus bavaricus*, Its Production in Pure Cultures, and Use in Lactic Acid Fermentation of Plant Material, German Patent 2,440,516, 1976.
55. Etchells, J. L. and Bell, T. A., Classification of yeasts from the fermentation of commercially brined cucumbers, *Farlowia*, 4, 87, 1950.
56. Etchells, J. L., Bell, T. A., Fleming, H. P., Kelling, R. E., and Thompson, R. L., Suggested procedure for the controlled fermentation of commercially brined pickling cucumbers — the use of starter cultures and reduction of carbon dioxide accumulation, *Pickle Pack. Sci.*, 3, 4, 1973.
57. Etchells, J. L., Borg, A. F., Kittel, I. D., Bell, T. A., and Fleming, H. P., Pure culture fermentation of green olives, *Appl. Microbiol.*, 14, 1027, 1966.
58. Chen, K. H., McFeeters, R. F., and Fleming, H. P., Stability of mannitol to *Lactobacillus plantarum* degradation in green beans fermented with *Lactobacillus cellobiosus, J. Food Sci.*, 48, 972, 1983.
59. Barnett, J. A., The utilization of sugars by yeasts, in *Advances in Carbohydrate Chemistry and Biochemistry*, Vol. 32, Pigman, W. and Horton, D., Eds., Academic Press, New York, 1976, 125.
60. Pederson, C. S. and Albury, M. N., The influence of salt and temperature on the microflora of sauerkraut fermentation, *Food Technol.*, 8, 1, 1954.
61. Fleming, H. P., Thompson, R. L., Etchells, J. L., Kelling, R. E., and Bell, T. A., Carbon dioxide production in the fermentation of brined cucumbers, *J. Food Sci.*, 38, 504, 1973.
62. McFeeters, R. F., Fleming, H. P., and Thompson, R. L., Malic and citric acids in pickling cucumbers, *J. Food Sci.*, 47, 1859, 1982.
63. McFeeters, R. F., Fleming, H. P., and Thompson, R. L., Malic acid as a source of carbon dioxide in cucumber juice fermentations, *J. Food Sci.*, 47, 1862, 1982.
64. Etchells, J. L., Fabian, F. W., and Jones, I. D., The *Aerobacter* fermentation of cucumbers during salting, *Mich. Agric. Exp. Stn. Tech. Bull.*, 200, 1945.
65. Fleming, H. P., Thompson, R. L., Etchells, J. L., Kelling, R. E., and Bell, T. A., Bloater formation in brined cucumbers fermented by *Lactobacillus plantarum, J. Food Sci.*, 38, 499, 1973.
66. Etchells, J. L., Borg, A. F., and Bell, T. A., Bloater formation by gas-forming lactic acid bacteria in cucumber fermentations, *Appl. Microbiol.*, 16, 1029, 1968.
67. Fleming, H. P. and Pharr, D. M., Mechanism for bloater formation in brined cucumbers, *J. Food Sci.*, 45, 1595, 1980.
68. Fleming, H. P., Etchells, J. L., Thompson, R. L., and Bell, T. A., Purging carbon dioxide from cucumber brines to reduce bloater damage, *J. Food Sci.*, 40, 1304, 1975.
69. Costilow, R. N., Bedford, C. L., Mingus, D., and Black, D., Purging of natural salt-stock brine for softening activity, *U. S. Dept. of Agric. Agric. Res. Serv.* ARS-72-5, 1977.
70. Daeschel, M. A., McFeeters, R. F., Fleming, H. P., Klaenhammer, T. R., and Sanozky, R. B., Mutation and selection of *Lactobacillus plantarum* strains that do not produce carbon dioxide from malate, *Appl. Environ. Microbiol.*, 47, 419, 1984.
71. Daeschel, M. A. and Fleming, H. P., Unpublished data.
72. Cruess, W. V. and Guthier, E. H., Bacterial decomposition of olives during pickling, *Univ. Calif. Agric. Exp. Stn. Bull.*, 368, 1923.
73. Taylor, S. L., Guthertz, L. S., Leatherwood, M., Tillman, F., and Lieber, E. R., Histamine production by food-borne bacterial species, *J. Food Safety*, 1, 173, 1978.
74. Mayer, K., Pause, G., and Vetsch, U., Fermentation characteristics and histamine formation in sauerkraut production, *Mitt. Geb. Lebensmittelunters. Hyg.*, 65, 234, 1974.
75. Taylor, S. L., Leatherwood, M., and Lieber, E. R., Histamine in sauerkraut, *J. Food Sci.*, 43, 1030, 1978.
76. Rice, S. L., Eitenmiller, R. R., and Koehler, P. E., Biologically active amines in food: a review, *J. Milk Food Technol.*, 39, 353, 1976.
77. Fleming, H. P., Etchells, J. L., and Bell, T. A., Vapor analysis of fermented Spanish-type green olives by gas chromatography, *J. Food Sci.*, 34, 419, 1969.

78. Aurand, L. W., Singleton, J. A., Bell, T. A., and Etchells, J. L., Identification of volatile constituents from pure culture fermentations of brined cucumbers, *J. Food Sci.*, 30, 288, 1965.
79. Bell, T. A., Turney, L. J. and Etchells, J. L., Influence of different organic acids on the firmness of fresh-pack pickles, *J. Food Sci.*, 37, 446, 1972.
80. Balatsouras, G., Tsibri, A., Dalles, T., and Doutsias, G., Effects of fermentation and its control on the sensory characteristics of Conservolea variety green olives, *Appl. Environ. Microbiol.*, 46, 68, 1983.
81. Porubcan, R. S. and Sellars, R. L., Lactic starter culture concentrates, in *Microbial Technology, Microbial Processes*, 2nd ed., Peppler, H. J. and Perlman, D., Eds., Academic Press, New York, 1979.
82. Lingle, M., Controlled fermentation of cucumbers, The "how's" and "why's", *Food Prod. Manage.*, July, 10, 1975.
83. Fleming, H. P., Thompson, R. L., Bell, T. A., and Hontz, L. H., Controlled fermentation of sliced cucumbers, *J. Food Sci.*, 43, 888, 1978.
84. Buescher, R. W., Hudson, J. M., and Adams, J. R., Inhibition of polygalacturonase softening of cucumber pickles by calcium chloride, *J. Food Sci.*, 44, 1786, 1979.
85. Raccach, M., Method for Fermenting Vegetables, U. S. Patent 4,342,786. 1982.
86. Daeschel, M. A. and Fleming, H. P., Entrance and growth of lactic acid bacteria in gas-exchanged, brined cucumbers, *Appl. Environ. Microbiol.*, 42, 1111, 1981.
87. Daeschel, M. A., Entrance of Microorganisms in Brined Cucumbers, Ph.D. thesis, North Carolina State University, Raleigh, 1982.
88. Fleming, H. P. and Etchells, J. L., Occurrence of an inhibitor of lactic acid bacteria in green olives, *Appl. Microbiol.*, 15, 1178, 1967.
89. Fleming, H. P., Walter, W. M., Jr., and Etchells, J. L., Isolation of a bacterial inhibitor from green olives, *Appl. Microbiol.*, 18, 856, 1969.
90. Fleming, H. P., Walter, W. M., Jr., and Etchells, J. L., Antimicrobial properties of oleuropein and products of its hydrolysis from green olives, *Appl. Microbiol.*, 26, 777, 1973.
91. Juven, B., Samish, Z., and Henis, Y., Identification of oleuropein as a natural inhibitor of lactic fermentation of green olives, *Isr. J. Agric. Res.*, 18, 137, 1968.
92. Garrido-Fernandez, A. and Vaughn, R. H., Utilization of oleuropein by microorganisms associated with olive fermentations, *Can. J. Microbiol.*, 24, 680, 1978.
93. Cruess, W. V. and Alsberg, C. L., The bitter glucoside of the olive, *J. Am. Chem. Soc.*, 46, 2115, 1934.
94. Juven, B. and Henis, Y., Studies on the antimicrobial activity of olive phenolic compounds, *J. Appl. Bacteriol.*, 33, 721, 1970.
95. Etchells, J. L., Costilow, R. N., Anderson, T. E., and Bell, T. A., Pure culture fermentation of brined cucumbers, *Appl. Microbiol.*, 12, 523, 1964.
96. Fleming, H. P., Etchells, J. L., and Costilow, R. N., Microbial inhibition by an isolate of *Pediococcus* from cucumber brines, *Appl. Microbiol.*, 30, 1040, 1975.
97. Sozzi, T., Watanabe, K., Stetter, K., and Smiley, M., Bacteriophages of the genus *Lactobacillus*, *Intervirology*, 16, 129, 1981.
98. Fleming, H. P., Humphries, E. G., and Macon, J. A., Progress on development of an anaerobic tank for brining of cucumbers, *Pickle Pack. Sci.*, 7, 3, 1983.
99. Gonzalez, C. F. and Kunka, B. S., Plasmid transfer in *Pediococcus* spp.: intergeneric and intrageneric transfer of pIP501, *Appl. Environ. Microbiol.*, 46, 81, 1983.
100. Klaenhammer, T. R., A general method for plasmid isolation in lactobacilli, *Curr. Microbiol.*, 10, 23, 1984.
101. Nes, I. F., Plasmid profiles of nine strains of *Lactobacillus plantarum*, in *Lactic Acid Bacteria in Foods* Proc. Neth. Soc. Microbiol., Meet., Wageningen, September 1983.
102. Stamer, J. R., Stoyla, B. O., and Dunckel, B. A., Growth rates and fermentation patterns of lactic acid bacteria associated with the sauerkraut fermentation, *J. Milk Food Technol.*, 8, 127, 1971.
103. Anon., Type Strains of *Lactobacillus* Species. A Report by the Taxonomic Subcommittee on Lactobacilli and Closely Related Organisms, American Type Culture Collection, Rockville, Md., 1968.
104. Starr, M. P., Stolp, H., Truper, H., Balows, A., and Schlegel, H. G., Eds., *The Prokaryotes. A Handbook on Habitats, Isolation and Identification of Bacteria*, Vol. 2, Springer-Verlag, New York, 1981, 1614.
105. Stamer, J. R., The lactic acid bacteria: microbes of diversity, *Food Technol.*, 33, 60, 1979.
106. Williams, R. A. D., A review of biochemical techniques in the classification of lactobacilli, in *Lactic Acid Bacteria in Beverages and Food*, Carr, J. G., Cutting, C. V., and Whiting, G. C., Eds., Academic Press, New York, 1975, 351.

107. Radler, F. and Yannissis, C., Weinsaureabbau bei Milchsaurebakterien, *Arch. Mikrobiol.*, 82, 219, 1972.

108. Jonsson, S., Clausen, E., and Raa, J., Amino acid degradation by a *Lactobacillus plantarum* strain from fish, *Syst. Appl. Microbiol.*, 4, 148, 1983.

Chapter 9

THE LACTOBACILLI AND STREPTOCOCCI: BAKERY PRODUCTS

T. Frank Sugihara

TABLE OF CONTENTS

I. HISTORICAL

The use of bacterial starters in the baking industry is still in its infancy. Recent developments here and abroad indicate a bright future for the use of pure culture bacterial starters in breads and other bakery products. With the sole exception of soda crackers, the "average" American diet does not include any bakery product using a bacterial fermentation. There are a few important limited products such as the "natural" sour rye bread that are available in various parts of the country, but these are by and large outnumbered by those made with chemical additives. Probably the most notable bakery product using a bacterial fermentation is the San Francisco sourdough French bread which accounts for approximately 20% of the bread consumed in a limited geographical area. The microorganism responsible for this unique bread is *Lactobacillus sanfrancisco.*[1-5] This unique bread is now being made in other geographical areas by the use of a pure culture bacterial starter.[6-8]

There are breads made in other parts of the world using bacterial fermentations that rely on the use of a natural "mother-sponge" as a starter. Some of these are the European sour ryes and Pumpernickel, the Shamsy and Kisra breads of the Arabic region, and the Idli and Puto breads of the South Pacific areas.[9-13]

Very important and complex bacterial fermentation systems are used in Italy for the production of sweet baked goods. They are Panettone, Colombia, Pandoro, and various mini-cakes using more or less the same dough formulations.[9,14]

The multimillion dollar soda cracker production in the U.S. still relies on naturally occurring microorganisms for fermentation. Recent studies showed that a pure culture bacterial starter system can be utilized to greatly shorten the process and also improve quality control.[15-17]

II. SPECIES INVOLVED IN BAKERY PRODUCTS

Over the years, a great number of bacteria have been isolated from fermenting dough systems but only a few have proven to be of commercial use.[9] This discussion will be limited to those being used commercially and to those that may have possible use in the future. Table 1 shows the important species isolated from bread doughs, sweet doughs, and soda cracker dough. The majority of the species listed are from the genus *Lactobacillus* and a few from *Streptococcus* and *Leuconostoc.* They all occur naturally on cereal grains, legumes, and animal products. When these products are mixed into a flour-water dough system and allowed to incubate at 25 to 35°C, they will metabolize the available carbohydrates into organic acids, diacetyl, ethanol, and carbon dioxide. These are the metabolites necessary to produce the flavor, aroma, and texture of various breads and other unique baked goods.

III. FUNCTIONS OF THE BACTERIAL STARTERS

In most cases, bacterial fermentation for bakery goods is quite similar in that it imparts a "distinctive" flavor and aroma to the final product. The famous San Francisco sourdough French bread involves a single bacterial species, *L. sanfrancisco,* which converts maltose from the flour into organic acids (primarily lactic and acetic) and other flavor compounds to produce its unique "bouquet". According to Schulz,[12] the lactic acid bacteria of major significance in sour rye fermentations are *L. plantarum, L. brevis,* and *L. fermenti.* The heterofermentative lactic acid formers are considered more important than the homofermentative types. Some of the homofermen tative lactic acid bacteria that produce lactic acid only as a rule are thermophilic bacteria whose temperature requirements cannot be maintained during sourdough fermen-

Table 1
BACTERIA ISOLATED FROM BAKERY FERMENTATION

Product	Species isolated
Bread	*Lactobacillus sanfrancisco*
	L. brevis
	L. plantarum
	L. fermentum
	Leuconostoc mesenteroides
	Streptococcus thermophilus
Sweet dough	*Lactobacillus brevis*
	L. plantarum
	L. fermentum
	Leuconostoc cremoris
	Streptococcus cremoris
Soda crackers	*Lactobacillus plantarum*
	L. delbrueckii
	L. leichmannii
	L. brevis
	L. casei
	L. fermentum

tation. On the other hand, acetic acid formation is an oxidative reaction, and thus the purely homofermentative acetic acid bacteria require an oxidizing medium. The carbon dioxide permeated substrate of the sourdough fails to provide a conducive environment for their activity. Lactic acid and acetic acid are, therefore, produced primarily by the heterofermentative bacteria of sourdough. *L. brevis*, a heterofermentative bacteria, also plays a major role in the fermentation of Italian sweet doughs used to produce Panettone, Colomba, Pandoro, and various mini-cakes.[9,14] A natural "mother-sponge" starter system has been used to initiate these fermentations for over 100 years. The fermentation involved in the process is a complex microbiological phenomenon. It involves a sourdough yeast, *Saccharomyces exiguus* (*Torulopsis holmii*), and a mixture of lactic bacteria including the dominant *L. brevis*. Here again, the major metabolites are the organic acids (primarily lactic and acetic acids).

Although the major function of bacterial fermentation in bakery goods is the formation of flavor and aroma, a few products rely on the microorganism for leavening and texture. Some of these products are Pumpernickel, Idli, and Puto breads. These products require a fairly long proof time to give sufficient time for carbon dioxide formation.

In all bakery products, the common substrate for fermentation is a cereal flour that contains sufficient sugars to supply the nutritional requirements of the lactic bacteria involved in producing metabolites such as lactic acid, acetic acid, ethanol, and diacetyl. In fact, most species of lactic bacteria can be grown in a water-flour system with or without the presence of a yeast. The metabolites from dough fermentation and baked goods have been described by many researchers.[18-23] Table 2 summarizes compounds identified in dough fermentation and baked goods.

IV. STARTER CULTURE PREPARATION

Interestingly, there are still many bakeries using the centuries-old technique of setting aside a small portion of the final fermented sponge for inoculating the next day's batch of sponge and dough. This is true with products such as sour rye bread, San Francisco sourdough bread, Italian Panettone, and Arabic Shamsy and Kisra. In most

Table 2
COMPOUNDS IDENTIFIED IN DOUGH FERMENTATION AND BAKED GOODS

Lactic acid[a]	Ethanol[a]	Diacetyl[a]
Acetic acid[a]	Propanol	Acetoin
Butyric acid	Isoamyl alcohol	Acetone
Propionic acid	Isobutanol	Acetaldehyde
Pyruvic acid	Isopropanol	Benzaldehyde
Formic acid	Amyl alcohol	Formaldehyde
Valeric acid	2,3-Butanedial	2-Hexanone
Caproic acid	β-Phenylethyl alcohol	Isovaleraldehyde
Caprylic acid	3-Methyl butanol	*n*-Valderaldehyde
Isocaproic acid	2-Methyl butanol	Methyethylketone
Lauric acid		Pyruvaldehyde
Myristic acid		*n*-Hexaldehyde
Hydrocinnamic acid		Isobutyraldhyde
Benzylic acid		Furfural
Succinic acid		3-Heptanone
n-Butyric acid		2-Methylethylbutaldehyde
Isobutyric acid		Hydroxymethylfurfural
Isovaleric acid		Phenolacetaldhyde
Heptanoic acid		2-Butanone
Pelagonic acid		Methional
Palmitic acid		Glyoxal
Orotonic acid		
Itaconic acid		
Leuvinic acid		

[a] Most abundant compounds. Other compounds found in trace quantities.

cases, a yeast such as *Saccharomyces exiguus* (*Torulopsis holmii*) or *Candida tropicalis* is part of the bacterial starter ecosystem. This technique has been very successful over the years since it is not only the source of souring activity but the leavening as well. The system is also self-preserving since the pH range is very acidic (PH 4.4 to 3.9) which helps prevent the growth of any contaminating microorganisms. A simple procedure for perpetuating a "mother-sponge" involves preparing a mixture containing 100 parts of the previous sponge, 100 parts of flour, and 46 to 52 parts of water. The resulting mixture is allowed to ferment for 8 hr at 30°C. It is then stored at 10°C for making up the next days' sponge. The main drawback to this system is the slow rate of leavening for today's large-scale commercial production schedule. However, this can be accelerated by the addition of Baker's yeast at the dough stage. The amount added can vary from 0.25 to 1% depending on the required time schedule.

The first "pure" bacterial starter culture for use in a bakery product was developed by Kline and Sugihara.[6-8] They developed a culturing media suitable for the propagation of *L. sanfrancisco*. This media was designated as SDB (sourdough bacteria) broth. The formulation is shown in Table 3. A solid medium can be prepared by adding 2% agar to the broth prior to sterilization. Cultures of *L. sanfrancisco* on SDB broth or SDB agar are incubated at 30°C for 1 to 2 days in an atmosphere containing a high level of carbon dioxide, attained by flushing the headspace above the broth cultures with CO_2, or in the case of the agar cultures, by placing the plates or slants in a receptacle wherein air is evacuated and replaced with a CO_2-air mixture containing 25 to 95 vol % of CO_2. After culturing, the cells can be harvested from the nutrient medium by centrifuging and washing with a 0.5% sodium chloride solution. The cells are then resuspended in a stabilizing menstrum consisting of 4 parts glycerol and 6 parts of an aqueous solution made up of 8% nonfat milk solids, 2% monosodium glutamate, and

Table 3
SOURDOUGH BACTERIA BROTH (SDB)

Ingredients	Grams
Maltose	2
Commercial yeast extract	0.3
Fresh yeast extractives[a]	0.5
Tween® 80 (sorbitan polyoxyethylene monooleate)	0.03
Trypticase (casein hydrolysate)	0.6
Water	100

Adjust to pH 5.6 with 20% lactic acid or with 1N to 6N HCl

[a] The fresh yeast extractives were prepared by autoclaving a 20% suspension of commerical Baker's yeast in distilled water for 30 min at 121°C, allowing the suspension to settle overnight at 2 to 8°C, decanting, and further clarifying the supernatant by centrifugation.

Data from Kline, L. and Sugihara, T. F., *Appl. Microbiol.*, 29, 459, 1971.

0.5% Tween® 80. The harvested cells are mixed with the stabilizing menstrum in a ratio of 1:4. The resulting stabilized cell concentrate can be frozen in liquid nitrogen and stored at −100°C (vapor phase of liquid nitrogen). It may be freeze-dried in suitable quantities. Both forms of the starter culture will retain full activity up to at least 12 months. Recently, a novel starter culture of *L. sanfrancisco* was developed by Kline.[8] A flour-water slurry was cultured by inoculating with an active sponge. The fully fermented slurry was then stabilized with 6% disaccharide and freeze-dried. The dried culture was then pulverized, packed in plastic bags, and stored at 0°C.

Many species of lactic bacteria can be propagated by use of a medium (MRS broth) developed by deMan et al.[24] The MRS medium is composed of the following: peptone, 10 g; dextrose, 10 g; yeast extract, 5 g; Tween® 80, 1 g; ammonium citrate, 2 g; sodium acetate, 5 g; magnesium sulfate, 0.1 g; manganese sulfate, 0.05 g; and disodium phosphate, 2 g/ℓ distilled water. The medium should be adjusted to pH 6.5 with hydrochloric acid. After culturing the bacteria in the MRS broth, the cells can be harvested, stabilized, frozen, or freeze-dried as described above for *L. sanfrancisco.* Procedures for preparing and storing frozen concentrated cultures of lactic starter bacteria are discussed and reviewed by Gilliland and Speck.[25]

For the production of soda crackers, Sugihara[16] found that a mixed starter culture of *L. plantarum, L. delbrueckii,* and *L. leichmannii* was necessary to obtain a satisfactory fermentation. The three species were grown separately, the cells harvested, and then mixed for a frozen concentrate or a freeze-dried starter culture. Conventional soda cracker fermentation, which relies on chance contamination for an inoculum, can be made more efficient by using pure-culture technology. Fermentation schedules would be much shorter and predictable. Quality control woud be greatly improved since delayed fermentations and "off-fermentations" would be eliminated. It would also be possible to control the flavor of the soda cracker by using different lactic bacteria or different combinations of lactic bacteria in the starter culture. Levels of organic acids as well as diacetyl can be controlled.

Recently, Hardy[26] developed a mixed culture starter for the production of sourdough bread in France. The medium for growing the starter consisted of enzyme (amylase)-

treated flour, 100 g; wheat germ, 5 g; ammonium sulfate, 13.4 g; potassium phosphate, 2.2 g; magnesium sulfate, 9.6 g; calcium chloride, 0.336 g; yeast extract, 1 g; and water, 1000 mℓ, pH 6.5. The inoculum consisted of *Candida tropicalis* and *L. plantarum* at 2×10^7/g. Incubation was at 30°C for 17 hr. The count for both species after incubation was approximately 2×10^9/g. The liquid mixed starter remained active for at least 3 weeks when stored at 4°C.

An interesting technique was devised by York[27] for making a sourdough bread at home. The starter used was a commercial yogurt that contained active *S. thermophilus.*

Conventional white bread is made with commercial Baker's yeast. The amazing fact is that much of the flavor of the bread is due to the lactic bacteria occurring as "contaminants" in the yeast cake. Baker's yeast contains, on the average, 5% lactic bacteria (range: 2 to 20%). If it were possible to control the type of lactic bacteria present, quality control of the white bread flavor would be greatly simplified.

As previously stated, the use of bacterial starter cultures in the baking industry is still in its infancy. Currently, the only bacterial starter produced and used in quantity is *L. sanfrancisco.* Four strains of *L. sanfrancisco* has been studied and found to have very little cytological and functional differences.[28] Soda cracker starter cultures are now available commercially, but the industry as a whole has not as yet accepted the technology. This is not surprising because since its development, more than 10 years passed before the adoption of pure culture technology for San Francisco sourdough French bread occurred.

The "mother-sponge" starter technique carried out for centuries is still being used for many bakery products such as sour rye bread, Pumpernickel, Panettone, Colomba, Pandoro, and others. Research is ongoing toward the development of pure bacterial starter cultures for some of these products. In the meantime the tedious "mother-sponge" starters will be used.

REFERENCES

1. Kline, L., Sugihara, T. F., and McCready, L. B., Nature of the San Francisco sour dough French bread process. I. Mechanics of the process, *Baker's Dig.*, 44, 48, 1970.
2. Sugihara, T. F., Kline, L., and McCready, L. B., Nature of the San Francisco sour dough French bread process. II. Microbiological aspects, *Baker's Dig.*, 44, 51, 1970.
3. Sugihara, T. F., Kline, L., and Miller, M. W., Microorganisms of the San Francisco sour dough bread process. I. Yeasts responsible for the leavening action, *Appl. Microbiol.*, 21, 456, 1971.
4. Kline, L. and Sugihara, T. F., Microorganisms of the San Francisco sour dough bread process. II. Isolation and characterization of undescribed bacterial species responsible for the souring activity, *Appl. Microbiol.*, 21, 459, 1971.
5. Sugihara, T. F. and Kline, L., Further studies on a growth medium for *Lactobacillus sanfrancisco, J. Food Protect.*, 38, 667, 1975.
6. Kline, L. and Sugihara, T. F., U. S. Patent No. 3,734,743, May 22, 1973.
7. Kline, L. and Sugihara, T. F., U. S. Patent No. 3,891,773, June 24, 1975.
8. Kline, L., U. S. Patent No. 4,243,687, January 6, 1981.
9. Sugihara, T. F., Non-traditional fermentations in the production of baked goods, *Baker's Dig.*, 51, 76, 1977.
10. El-Gendy, S. M., Fermented foods of Egypt and the middle east, *J. Food Protect.*, 46, 358, 1983.
11. Mukherjee, S. K., Albury, M. N., Pederson, C. S., Van Veen, A. G., and Steinkraus, K. H., Role of *Leuconostoc mesenteroides* in leavening the batter of Idli, a fermented food of India, *Appl. Microbiol.*, 13, 227, 1965.
12. Schulz, A., Fundamentals of rye bread production, *Baker's Dig.*, 40, 77, 1966.
13. Meyer, S. J., Pan rye bread production, *Baker's Dig.*, 39, 44, 1965.
14. Zorzanello, D. and Sugihara, T. F., The technology of pandoro production, *Baker's Dig.*, 56, 12, 1982.

15. Sugihara, T. F., Microbiology of the soda cracker process. I. Isolation and identification of microflora, *J. Food Protect.*, 41, 977, 1978.
16. Sugihara, T. F., Microbiology of the soda cracker process. II. Pure culture fermentation studies, *J. Food Protect.*, 41, 980, 1978.
17. Sugihara, T. F., U. S. Patent No. 4,353,926, October 12, 1982.
18. Johnson, A. H., Identification and estimation of the organic acids produced during bread dough and cracker dough fermentation, *Cereal Chem.*, 2, 345, 1925.
19. Ng, H., Factors affecting the organic acid production of sour dough (San Francisco) bacteria, *Appl. Microbiol.*, 23, 1153, 1972.
20. Johnson, J. A. and Sanchez, C. R. S., The nature of bread flavor, *Baker's Dig.*, 47, 48, 1973.
21. Magoffin, C. D. and Hoseney, R. C., A review of fermentation, *Baker's Dig.*, 48, 22, 1974.
22. Peterson, C. S., *Microbiology of Food Fermentation*, AVI Publishing, Westport, Conn., 1971, chap. 8.
23. Jackel, S. S., Fermentation flavors of white bread, *Baker's Dig.*, 43, 24, 1969.
24. de Man, J. C., Rogosa, M., and Sharpe, M. E., A medium for the cultivation of lactobacilli, *J. Appl. Bacteriol.*, 23, 130, 1960.
25. Gilliland, S. E. and Speck, M. L., Frozen concentrated cultures of lactic starter bacteria. A review, *J. Milk Food Technol.*, 37, 107, 1974.
26. Hardy, J. L., Mise au d'une Technique de Panification sur Levain a Partir de Ferments Cerealiers Associes, Docteur Ingenieur These, Universite de Technoligie de Compiègne, Compiègne, France, 1982.
27. York, G. K., The sourdough breakthrough, *Sunset Mag.*, May, p. 112, 1973.
28. Sriranganathan, N., Seidler, R. J., Sandine, W. E., and Elliker, P. R., Cytological and deoxyribonucleic acid-deoxyrobonucleic acid hybridization studies on *Lactabacillus* isolates from San Francisco sourdough, *Appl. Micobiol.*, 25, 561, 1973.

Chapter 10

TYPES OF STARTER CULTURES

Randall K. Thunell and William E. Sandine

TABLE OF CONTENTS

I. INTRODUCTION

While fermented milks date back into antiquity, the discovery and characterization of the causative bacteria have occurred more recently. Originally, fermented milks resulted from natural contaminating microorganisms present in raw milk or residing in the cracks and crevices of milk containers. The practice of inoculating milk with a pure culture of starter bacteria was a development of the early 1900s. Prior to this time, cultures were carried in much the same way as sourdough cultures are carried today, i.e., a portion of the old "sour" was used to inoculate fresh dough.

Modern lactic starter technology is a highly competitive, multimillion-dollar sector of today's cheese industry. More literature exists on milk fermentative organisms than for any other food. Several excellent review articles have been written detailing the history of starter cultures.[1-3]

This chapter will discuss the use of the homofermentative, mesophilic lactic streptococci *Streptococcus lactis* and *Streptococcus cremoris* (now considered a subspecies of *S. lactis*)[4] in hard and soft cheesemaking. Also discussed are thermophilic lactic starters. These starter cultures serve several important functions in the cheesemaking process including initiation of the fermentation process, inhibition of potentially pathogenic organisms, enhancement of cheese keeping quality, direct and indirect contributions to cheese flavor and aroma, and the stabilization of milk in the form of cheese.[5]

II. MESOPHILIC STARTERS

A. Strain Interactions and Dominance

1. Phage

Successful cheesemaking relies on the conversion of milk sugars to lactic acid by starter bacteria. It is uncontested that bacteriophage (phage) infection of starter cultures represents the most challenging problem to the cheese industry, as well as the greatest source of economic loss. Cheesemakers, in the face of this dilemma, have had only two problem-solving alternatives. The first alternative is to practice some form of asepsis in attempting to prevent phage contamination. State-of-the-art procedures for this include propagation of strains under aseptic conditions, the rotation of phage-unrelated strains, use of phage-inhibitory media, sanitation of vats and utensils, early renneting and elimination of ripening time, addition of higher levels of inoculum, and the use of specially constructed starter tanks that prohibit phage entry.[1,6,7] According to Hull,[8] asepsis constitutes an artificial system of control and suffers from the disadvantage that the starter remains sensitive to infecting phages in the plant and eventually will be adversely affected when phage contamination occurs.

The second alternative allows or even encourages phage contamination of the mother or bulk starter cultures. The contaminating phages challenge bacterial cells, killing the sensitive types and allowing the growth of phage-resistant bacteria.[1,9,10] In this regard, the practice of adding virus-laden whey to a multiple-strain starter culture for acid-casein manufacture has been successfully used in both New Zealand[9] and the Netherlands.[11] New Zealand researchers reported[9] that acid production from a multiple-strain starter culture, without added whey, was initially satisfactory, but began to slow down after 10 days. The addition of phage-bearing whey, which subjected the culture to continuous selection pressure, extended the useful life of the culture to over 1 month.

Sandine[6] found that a single composite whey sample, when tested against 64 different cheese starter-derived single strains, revealed phages in the cheese factory enviroment for almost half of the strains. Hull[8] took advantage of the prevalence of viruses in cheese factories in using natural selection pressures to adapt specific starters to a plant environment. When active phages were detected, the corresponding starter culture was adapted by adding whey, thereby building into the starter a natural resistance to the phage population in the factory.

Starter "recovery", after a phage attack, not arising from the appearance of phage-resistant mutants, has also been reported in the literature.[12] In this case a dominant strain in a culture was attacked and destroyed by phage. The remaining strain(s), though once minor in percentage population of the culture, became dominant through subculturing.

In an effort to minimize phage infection, cheesemakers have resorted to rotating several different mixed-strain (undefined-strain blend) cultures. The theory behind this practice suggests that each culture contains different bacterial strains that are theoretically insensitive to the same infecting phage. Thus, by rotating cultures during cheesemaking, an indicator strain (one capable of supporting phage growth) will presumably not be in use long enough to support phage proliferation with concomitantly impaired acid development.

Mixed-strain cultures, for the most part, are undefined blends of strains of *S. lactis* and *S. cremoris* that can vary in their acid-producing activity, flavor characteristics, and phage sensitivities; their use results in a finished cheese of varying quality and flavor. Often cheesemaking processes must be altered to accommodate the properties of each mixed-strain culture. It is not uncommon to find mixed-strain cultures from different culture supply houses being used in large rotation schemes.[13,14] It is likely that some of these cultures are identical or closely related in strain composition, thus creating a precarious starter culture program in light of research findings that have shown starter bacteria themselves to be the most likely source of infecting bacteriophages.[6,15,16]

It is well documented that lactic streptococci found in commercial starter cultures exhibit lysogeny (the presence of prophage in the bacterial genome) and are inducible.[17-20] This partially explains why it is difficult to achieve bacteriophage-free conditions in a cheese plant. Cheese factories using large rotation schemes can usually expect to encounter phage problems since the presence of indicator strains as well as phage-related strains is highly probable. Lawrence[17] noted that the use of several supposedly different mixed-strain starters greatly increased the risk of bacteriophage infection due to the fact that more temperate phages were released and more indicator strains were present to support phage replication. It was also noted that the greater the number of starter cultures in use, the greater the opportunity for undesirable phage modifications and mutations to occur. The possibility that unknown phages arise as virulent mutants of temperate phages may account for the sudden appearance of phages virulent for newly isolated starter strains.[21] Bacteriophage can be effectively eliminated from the cheese plant environment by removing the indicator strains upon which they proliferate.[15,22] This presents a formidable obstacle, however, when working with culture rotations of unknown starter blends, since accurate phage monitoring requires that phage or whey samples be assayed against individual strains. Also, research conducted by Sanders and Klaenhammer[23] showed that the elevated temperatures encountered during normal Cheddar cheesemaking can result in temporary loss or reduction in the restriction-modification activities of starter bacteria, thus leading to the appearance of lytic bacteriophages which are modified for efficient multiplication on a new host.

Phage-carrying (pseudolysogenic) cultures represent another source of bacteriophage contamination.[14,24] Such cultures will continue to grow in the presence of phage which are being propagated by a nonproductive, phage-sensitive strain in the culture. This nonproductive strain can be the means of propagating and distributing phage and this can be lytic for susceptible strains of greater importance to other cultures in rotation. It has also been found that a small, but constant proportion of a starter culture can support phage growth through many culture transfers.[24]

Gasson and Davies[25] were successful in curing strains of *S. lactis* and *S. cremoris* of prophage with ultraviolet irradiation. They suggested that future strain improvement programs may well include the process of curing prophage from commercially useful starter cultures.

2. Subculturing

Heterogeneity among strains of lactic streptococci is now commonly recognized, and the single strains may contain variant cells that differ greatly in their phage sensitivities, phage adsorption patterns, phenotypic expression, and host modification-restriction systems.[17,26] It is not uncommon for the proportions of these strain-variant cells to change on subculturing.[15] Limsowtin et al.[26] have shown that the appearance of strain variants was dependent upon the nature of the subculturing medium. Cultures propagated and transferred in enriched broth media such as M17[19] accumulated variants more rapidly than when propagated in sterile skim milk. When maintained routinely in the laboratory, many strains readily change their phenotypes. Gilliland[27] found that strain balances differed for cultures grown in a broth medium with the pH controlled at 6.0 as compared to the same culture grown in milk. However, the composition of the growth medium was found to be responsible for altering the strain balance, rather than pH-controlled conditions.

Temperature has an important influence upon culture end products and growth.[28,29] Lee and Collins[28] examined the optimal and maximal growth temperatures for nine strains of *S. cremoris.* Optimal temperatures in yeast extract broth varied from 28 to 31°C, while maximal temperatures ranged from 34.4 to 38.9°C. These data emphasize that even small temperature shifts while subculturing can predispose one strain to dominate others. These domination effects can be further exaggerated by the growth medium selected.

Collins[30] has attributed competitive growth abilities to domination during subculturing. These would include small differences in lag phase and growth rate, differences in tolerance to end products, variations in nutritional requirements, physiological state of the organism, and production of antibiotics by individual strains, e.g., nisin.

3. Bacteriocin Production

The potential of lactic streptococci to produce metabolites that are inhibitory to other bacteria of the same species and to other genera, is well documented.[30-35] Collins[30] reported that 107 of 110 culture mixtures were overcome and dominated by antibiotic-producing cultures in only 1 to 2 days. Domination of a defined-strain culture by a bacteriocin-producing strain also has been observed by the authors under actual cheesemaking conditions. Such domination was followed by phage attack on the dominating strain, resulting in starter failure. According to Geis et al.[33] frequency of bacteriocin production is approximately 9% in strains of *S. lactis* and 7.5% in strains of *S. cremoris.* Davey[36] has shown that diplococcin, the bacteriocin produced by strains of *S. cremoris,* rapidly stopped DNA and RNA synthesis and greatly reduced protein synthesis. In view of these findings, the importance of selecting nonbacteriocin-producing strains cannot be overlooked without disastrous results.

The nature of the "Nascent Phenomenon" and "Lysis-from-Without" is related to a phage-induced bacteriocin called lysin.[37-40] This bacteriocin is coded by the genes of a virulent phage within the host bacterium and causes the host cell to burst and release mature phage particles. The released lysin, however, is nonspecific and is capable of lysing a wide variety of starter strains. Tourville and Johnstone[38] found that lysin released upon phage lysis of *S. lactis* C10, caused lysis of all lactic streptococci examined.

4. Stimulatory Compounds

The rate of lactic acid production by a single-strain starter can often be increased when the strain is grown in association with another starter strain. Improved acid production was attributed to one of the cultures being stimulated by metabolic by-products released by the other organism.[40-42] Kothari et al.[43] examined the associative growth of the starter organisms by measuring their acid production when grown both singly and as pairs in milk. Of the strains examined, 13 pairs of cultures showed increased acid production as compared to acid production by the single strains. Dahiya and Speck[42] identified adenine as the main stimulatory compound released into milk during associative growth. They also showed that some strains are unable to synthesize required amounts of adenine for optimal growth and that growth could be stimulated by the addition of adenine to the culture medium. The pairing of single-strain starters tends to lessen seasonal changes in activity, and produces more consistent activity over the course of the year.[2]

B. Defined-Strain Starters

1. Historical Development

The use of single strains of lactic streptococci for Cheddar cheesemaking originated in New Zealand[25] with Whitehead. The New Zealand concept of rotating three or four pairs of phage-unrelated single strains remained the same for almost 40 years. Strain pairs usually consisted of one strain sensitive to cooking temperatures (38 to 40°C), paired with a strain somewhat insensitive to these temperatures. In 1976, three pairs of strains were combined into a multiple blend starter.[44,45] After 2 years of use, the decision was made to again rotate 3 pairs of strains. In 1980, a single pair of phage-insensitive strains was introduced. Since then most New Zealand cheese plants have continuously used this same pair of strains.[44]

The Australian defined-strain system involves the use of several single strains in rotation. As phage appear for a strain in the rotation, it is replaced by a phage-insensitive mutant derived from the infected strain.[7,17] During the past 20 years, many small cheese producers have merged to form large cooperative and independent cheese plants. Starter cultures are placed under tremendous stress in these large cheese plants where multiple-vat fills are common, and daily milk processing often exceeds 2 million pounds. Such practices require shorter, more scheduled manufacturing times. These conditions tend to accentuate variations in starter performance. Also, as mentioned above, mixed-strain starters of unknown composition tend to show greater variability from culture to culture, often requiring modifications in manufacturing procedures to compensate for culture characteristics.

Advantages attributable to defined-strain cultures include the continuous use of the same culture every day, more consistent and predictable starter activity, no culture rotations, elimination of starter failure due to phage, lower inoculum levels, less seasonal effects, upgraded cheese quality, standardized production conditions, and the ability to monitor individual strain characteristics (e.g., phage sensitivity, flavor, acid production, compatibility, agglutination, and lysogeny).[2,17,22,44-50] The production of consistent and predictable starter activity for cottage cheesemaking is demonstrated in Figure 1. The same defined-strain starter was used in all 100 vats of cottage cheese. Figures 2 and 3 compare pH ranges of Cheddar cheese after 24 hr. The pH values for Cheddar cheese manufactured with defined-strain cultures fell within a much narrower range than those of cheeses made with commercial mixed-strain cultures. The economic ramifications of these data have been discussed in futher detail by Thunell et al.[48]

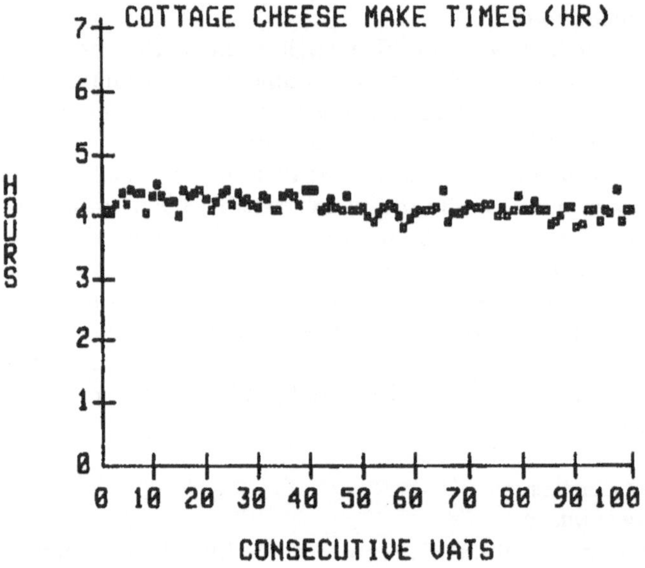

FIGURE 1. Uniformity in make time for sequential vats of cottage cheese made within the same defined strain culture for 100 consecutive days. (From Thunell, R. K., Sandine, W. E., and Bodyfelt, F. W., *J. Dairy Sci.*, 67, 1175, 1984. With permission.)

2. Selection of Strains

Starter strains presently in use, as well as replacement strains, are subjected to a formidable range of lytic bacteriophages that normally exists in the cheese plant environment. It thus becomes imperative that the selection of starter strains include those that are insensitive to phages in the plant environment or show very low phage multiplication factors.[17] It should be reemphasized that a plant that uses a large number of strains, even defined strains, greatly increases chances for phage problems because of the increased likelihood of strain relatedness.[51]

Several tests have been devised to aid in the selection of suitable starter bacteria. The test of Horrall and Elliker[52] categorized starter cultures as being slow, medium, or fast acid-producers at 37°C over a cheesemaking interval. Use of this temperature, however, likely discriminates against useful strains of *S. cremoris*. These, desirably so, are inhibited by Cheddar cheese cooking temperatures.[28] Pearce[53] developed a test (New Zealand Activity Test) to predict the acid-producing abilities of starter cultures. This test subjected strains to the same profile of times and temperatures that are encountered by starters in the cheese vat. Recently, Heap and Lawrence[54] modified this test to improve the relationship between laboratory results and starter activity in the cheese vat. This was done by adjusting culture inocula to the same densities before the start of an activity test.

Richardson et al.[55] described a simple laboratory activity test to determine the presence of inhibiting phage. New strains were screened for phage sensitivity prior to their introduction into cheesemaking using the test of Heap and Lawrence.[56] Using this test, the existence or development of a phage that was active against a potential starter strain could be anticipated before the strain was actually used in cheesemaking. Some attempt also has been made to select starter strains by computer, based on a "resistance index" of potential strains to various races of bacteriophage.[57,58]

Bitterness, as a flavor defect in cheese, has been linked directly to bacteria that survive cook temperatures and continue proliferating to high cell numbers in the

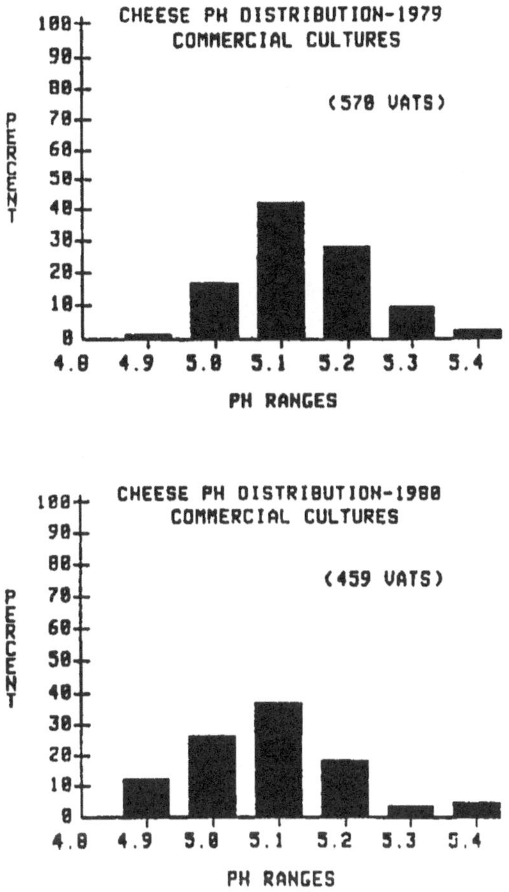

FIGURE 2. Broad pH range for Cheddar cheese made with rotated commercial mixed strain starters of unknown compostion. (From Thunell, R. K., Bodyfelt, F. W., and Sandine, W. E., *J. Dairy Sci.*, 67, 1061, 1984. With permission.)

curd.[58,60-65] Sullivan et al.[59] devised a simple laboratory method for classifying starter strains as potentially bitter or nonbitter according to their growth in milk at normal cooking temperatures (37 to 38°C). The viability of cultures indicated their potential to produce bitter flavors in cheese. The most recent hypothesis concerning bitterness suggested that all strains are potentially capable of producing bitter flavors in cheese in response to selected manufacturing procedures.[63] Any manufacturing conditions that permitted starters to multiply to high numbers in the curd resulted in bitter cheese. Most defined-strain cultures used today contain strains of *S. cremoris* rather than *S. lactis* because the former exhibit greater sensitivities to cook temperatures and tend to produce better flavors.[51] New Zealand researchers have recommended that temperature-sensitive and -insensitive strains be mixed in a 2:1 ratio, respectively, to prevent bitterness.[45] In the U.S., however, we have found this step unnecessary. Cheese free of bitter flavors has been manufactured in the U.S. with defined strains for over 5 years without attempts to adjust ratios of strain or temperature sensitivities.[22]

3. Phage-Insensitive Mutants

The first attempts to derive phage-resistant mutants from phage-challenged cultures

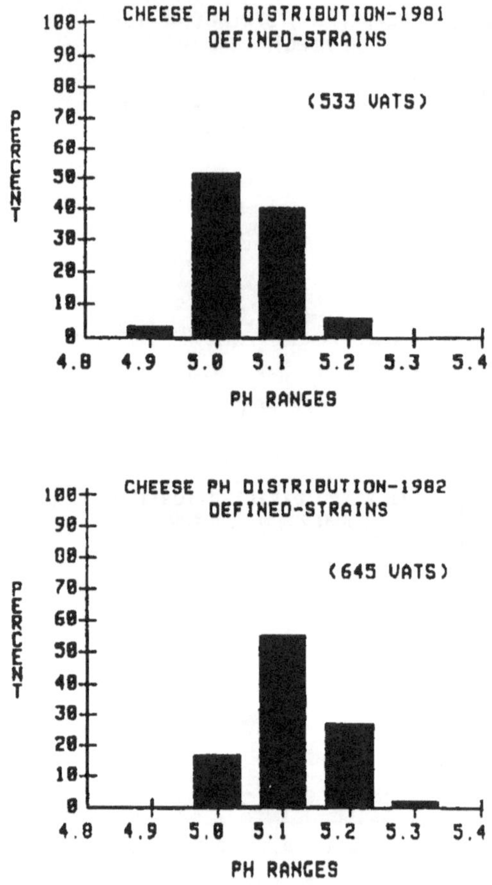

FIGURE 3. Narrow pH range for Cheddar cheese made with unrotated defined strain starters of known composition. (From Thunell, R. K., Bodyfelt, F. W., and Sandine, W. E., *J. Dairy Sci.*, 67, 1061, 1984. With permission.)

date back to Whitehead in 1936.[66] Also, bacteriophage-insensitive mutants have been used in Cheddar cheese manufacture for approximately 10 years with varied success.[67-70] Several possible reasons have been given to account for phage insensitivity.[69] These include altered receptor sites on the cell surface, the action of a restriction/modification system on invading phage DNA, failure of the phage particle to inject its DNA, and lysogenic immunity due to the presence of a prophage.

Most mutant-associated difficulties have centered around reversion of the new strain to original phage sensitivity, increased host ranges for phage, sensitivity to new phages, and insufficient acid production.[24,68,69] Limsowtin and Terzaghi[69] speculated that some of the early attempts to select phage-insensitive mutants may have failed because clear differentiation was not made between genotypic phage-resistance and the phage-carrying state, which can manifest phenotypic resistance. The effectiveness of phage monitoring is dependent upon both the indicator strain selected, and upon the culture medium used for its propagation.[26] As mentioned above, indicator strains grown in enriched broth may not be the same as those grown in milk.

Most procedures for the isolation of phage-insensitive mutants usually involve some form of whey adaptation (the addition of phage-laden whey to an actively growing

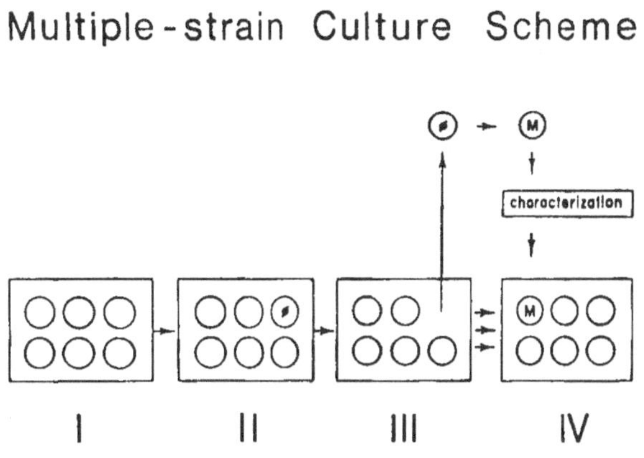

FIGURE 4. Scheme showing the phage-resistant strain-replacement technique used in the Oregon State University defined-strain culture program.

culture).[7-10,22,67,69,70] Once phage-insensitive cells have grown, they are plated on a differential medium for selection of fast acid-producing clones[71] (see Chapter 2). Other common sources of phage-insensitive mutants include phage-carrying cultures, or commercial mixed-strain cultures, where resistant strains can be selected through single-colony isolation techniques.[24]

In spite of difficulties encountered with some phage-insensitive mutants, at least two programs for selection of phage-insensitive mutants are still successful after several years of operation. An Australian starter system[67] built around the use of phage-insensitive mutants has eliminated failures caused by phage attack for more than 3 years.

A phage-insensitive, multiple-strain approach was developed at Oregon State University in the late 1970s, which has since been adopted by several culture suppliers in the U.S.[22,49] This system centers around the selection of a multiple-strain blend of two to six strains of *S. cremoris*. These strains are initially insensitive to phage present in a plant environment. Over the course of time, as phage appear or adapt for strains in the blend, the infected strains are removed and cheesemaking continues normally using the other strains in the blend (Figure 4). The infected strain is subjected to whey adaptation and then plated on fast-slow differential agar (FSDA, see Chapter 2). Fast acid-producing colonies are picked and characterized for desirable parental traits, and then are returned to the multiple-strain blend for cheesemaking. Mutant-strain stability in the Oregon State University system has been remarkably good as shown in Table 1. Some mutant strains have been used in cheesemaking for almost 4 years.

Coupled with the selection of phage-insensitive mutants is an ongoing program of phage monitoring in cheese plants using this defined-strain system. It should be noted that the presence of phage for a strain does not warrant its immediate removal from a culture blend. Many times phage will appear and cycle with low multiplication factors for a strain without seriously impairing its acid-producing function. Phage monitoring helps to anticipate a critical buildup of phage for strains in the blend beyond this low-level cycling.

4. *Proteinase-Negative Strains*

Upon original isolation and propagation, most starters contain essentially all proteinase-positive (prt^+) cells. However, daily transfer of a culture over a long period of time results in an accumulation of proteinase-negative (prt^-) or proteinase-deficient cells.[72] Proteinase-negative cells appear at a relatively high frequency (1 to 2%) and

Table 1
MUTANT STABILITY REPRESENTED BY MONTHS USED IN COMMERCIAL CHEESEMAKING

	Months used in cheesemaking		
Mutant	Plant A	Plant B	Plant C
290-PC	33[a]	43[a]	43[a]
M290-PC	6	—	—
108-12	—	45[a]	—
108-1	—	3	—
108-C	30[a]	—	—
SKII-W	9	—	—
SKIIG-C	36[a]	36[a]	37[a]
SKIIG-D	—	10	—
224-3	51[a]	54[a]	44[a]
C13/25	30[a]	39[a]	40[a]
U-134[b]	36[a]	36[a]	36[a]

[a] Still in cheesemaking use.
[b] Phage-insensitive mutant derived in New Zealand from *S. cremoris* strain AM2.

slow acid production eventually results.[6] The limited growth of proteinase-negative strains in milk is due to low surface-bound proteinase activity.[73] Lactic streptococci are fastidious in their nutritional requirements.[74] Richardson[72] has shown that prt$^-$ variants will continue to accumulate in a culture until their protein demand cannot be met by the medium or by the action of prt$^+$ organisms, at which point they cease multiplying. Mills and Thomas[75] studied the effects of levels of starter proteinase on bitterness development in Cheddar cheese. The level of starter proteinase was varied by replacing varied amounts of prt$^+$ with prt$^-$ organisms. Total cell densities remained constant. Findings showed that cheese made with 45 to 75% prt$^-$ cells developed less bitterness than cheeses made with prt$^+$ strains alone. Richardson et al.[76] reported that high concentrations of prt$^-$ variants could produce sufficient acid in milk over a 5-hr period to qualify them for cheesemaking use. Generation times for prt$^-$ strains were 3.5 times longer in cheese milk than those for prt$^+$ organisms. Because bacteriophage require actively dividing cells for replication, prt$^-$ strains offer a possible means of helping control bacteriophage. Antibiotics that affect cell division would also exhibit less inhibition on prt$^-$ strains. Studies are presently underway[72] to determine if significant increases in yield can be realized through the use of the less-proteolytic strains. Shelaih et al.[77] examined the temperature sensitivities of prt$^-$ variants and found that with increased inoculum rates, prt$^-$ cells would produce more acid at cook temperatures than comparable prt$^+$ strains, thus allowing for cheesemaking at higher temperatures.

Advantages of prt$^-$ strains are summarized as follows: cell growth has been effectively separated from acid production (i.e., growth occurs in the bulk tank and only slightly in the cheese vat due to longer generation times), acid production is consistent in the cheese vat, less casein is proteolyzed, bacteriophage problems are eliminated, and chances for bitter flavor production are reduced.[72]

C. pH-Control Systems

A 1973 New Zealand survey of bulk starter preparation and handling procedures revealed that in many cases irregular acid production resulted from the use of overripe bulk starter.[78] Harvey[79] reported that growth of *S. lactis* at pH values less than 5.0

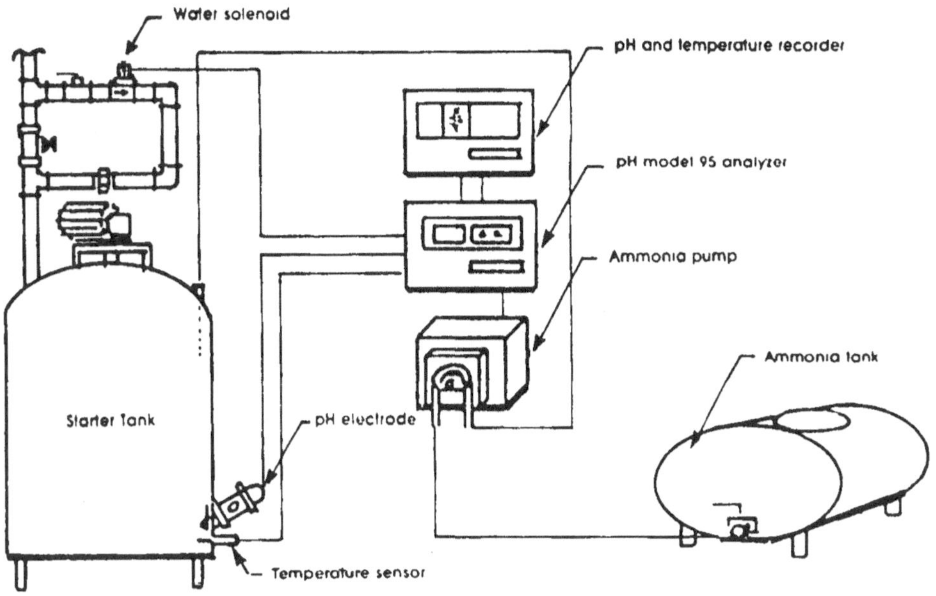

FIGURE 5. Diagram showing external pH and temperature control systems for a bulk starter tank. (Courtesy of Biolac Inc., Logan, Utah.)

resulted in damage to cellular enzyme systems. Other experiments showed that bulk starter progressively lost activity when held for periods below pH 4.7.[78] Of the two species most commonly used in starter cultures, strains of *S. cremoris* were generally more acid sensitive than strains of *S. lactis.* The last decade has brought significant contributions to the field of starter technology, particularly in the area of pH-control for bulk starter preparation.

1. External Control

In 1977, Richardson et al.[80] reported on the development of a pH-controlled bulk starter system for Cheddar cheese manufacture. Whey was fortified with phosphates and yeast extract to provide an enriched culture medium. The pH of the medium was maintained at 6.0 during bacterial growth by injections of anhydrous or aqueous ammonia. Culture tanks were fitted with electrodes that could withstand pasteurization, and an analyzer-recorder instrument continuously monitored acid production and neutralization in the growing culture (Figure 5).[81] The system resulted in a starter culture with acid-producing activity superior to that of conventional phage-inhibitory or milk media. In addition, inoculum levels were reduced by 70 to 80% and the starter could be held over without overripening; acid production ceased because lactose was completely utilized by the culture during growth. The application of pH-control also allowed for the reduction of phosphates in the medium without compromising phage-inhibitory benefits.[82] Since its introduction, the pH-controlled whey-base system also has been successfully used in the production of cottage, Italian, and Swiss cheese cultures.[83,84]

Sophisticated pH/temperature-control instrumentation using microprocessors is now available from several U.S. manufacturers. Phage-inhibitory media for use with pH-control systems now range from a simple growth supplement added to whey to complete milk-based media. Where equipment costs are prohibitive, a manual two-step neutralization procedure for preparing bulk starter has been described.[85] External pH-control offers the lowest-cost bulk starter available. Major savings, of up to two thirds in media costs, have been reported with pH-control systems.[86]

Growing starter bacteria under optimal pH and temperature conditions, coupled with the stimulatory effects of added yeast extract, has allowed strains that normally exhibit slower rates of acid production to be used in cheesemaking. Inoculum levels of these cultures average 0.5 to 0.75%, with 3.5- to 4.0-hr make times. Ripening periods also have been eliminated in most cases.[22] Slower acid-producing strains are generally less susceptible to phage attack than are fast acid-producing strains.[51]

2. *Internal Control*

In 1979, researchers at Oregon State University announced the development of an internal pH-controlled medium that relied on the solubilization of buffers contained in the medium in response to acid production by the growing cultures.[87,88] In 1981, a U.S. patent was issued for this unique starter medium.[89] Internal pH-control offers many of the advantages of external control, but requires no equipment investment or neutralizing chemicals. Some of the advantages noted for internal pH-control include improved holding characteristics, no overripe starter, lower inoculum levels, phage neutralization, greater starter activity, and built-in pH-control.[88]

D. Conclusions

The two most important improvements in culture technology in the last decade have been the application of defined-strain starter cultures to U.S. cheesemaking, and the growth of starter bacteria under pH-control conditions. It is likely that further research in these areas will center around the selection of particular strains for improved flavor characteristics, the genetic stabilization of strains against phage, and yield improvements through the commercial application of proteinase-negative strains in cheesemaking.

III. THERMOPHILIC STARTERS

Thermophilic lactic starter cultures, also known as rod-coccus cultures, are made up of a rod-shaped bacterium, usually *Lactobacillus bulgaricus, L. lactis,* and/or *L. helveticus* and the coccus-shaped bacterium *Streptococcus thermophilus.* Mixed-strain starters of this type, with an optimum termperature around 40°C, are used in the manufacture of yogurt and related fermented milks and Swiss- and Italian-type cheeses. The Swiss types include Emmentaler, Gruyère, and Swiss cheese while the Italian types include Mozzarella, Pizza, Grana (Parmesan and Romano), Provolone, Gorgonzola, and Caciocavallo cheeses.

While the Italian- and Swiss- style cheeses may rely on the same organisms in their manufacture, the ripening processes vary considerably and thus the final products are dissimilar. Another difference is that Swiss-style cheese contains strains of *Propionibacterium* that produce CO_2 during the warm-room treatment. Both of these cheeses were originally made in Italy and Switzerland using naturally soured raw milk. Details on their manufacture have been published by Kosikowski[90] and Reinbold.[91,92]

According to Wood[93] the practice in Italy is to use whey starters for the manufacture of the Italian cheese that is made only from raw milk. Whey is taken from the vat shortly after the cooking of the curd and is pumped to starter tanks without additional heat treatment. The temperature of the whey is 45 to 55°C and it is held undisturbed for about 24 hr until the desired titratable acidity (1.12 to 1.18%) is reached. Italian law prohibits the use of bacteriophage-inhibitory media for preparing bulk starter cultures. The law further requires that only fresh milk be used for starter growth, with the exception that whey can be used for whey starters. Nonfat dry milk solids may not be used in starter media.

Accolas and Auclair[94] and Auclair and Accolas[95] have recently discussed the taxonomy of *S. thermophilus, L. bulgaricus, L. lactis,* and *L. helveticus.* Table 2 summarizes

Table 2
CHARACTERISTICS OF THERMOPHILIC BACTERIA USED IN LACTIC STARTERS

Test	*S. thermophilus*	*L. bulgaricus-lactis*	*L. helveticus*
Morphology	Cocci, pairs & chains	Rods, variable with age	Rods, less polymorphic
Metachromatic granules	—	Yes	No
Lactic acid isomer		D	DL
Growth at 45°C	Yes	Yes	Yes
Growth at 15°C	No	No	No
Growth at 10°C	No	No	No
% Acid in milk	1.8	<1.8%	>2%
NH_3 and CO_2 from urea	Yes	No	No
Ferments maltose	No	Yes	Yes
Survives 65°C, 30 min	Yes	—	—
G + C %	40	50	39

the main properties of these organisms. In considering the composition of thermophilic starters, these authors distinguish two types. First there are the artisan starters with a variable or unknown composition. These are in common use in small cheese factories in the Jura and Alps of Switzerland where Gruyère and Emmentaler cheeses are made. These starters are produced by grinding previously dried calf stomachs in whey saved from the cheese vat at the end of each working day. After deproteinizing by heating and acidifying at 40 to 45°C, the starter is used and is not only a source of thermophilic lactic bacteria, but of milk-coagulating enzyme as well. Bacteria commonly found in such artisan starters include *L. fermentum, L. helveticus, L. lactis, L. bulgaricus,* and *L. acidophilus,* in addition to *S. thermophilus* and occasionally fecal streptococci of the enterococcus group.

The second type of starter described by Accolas and Auclair is of known composition. Selected strains of *S. thermophilus* and lactobactilli (*L. bulgaricus* and *L. helveticus*) are used in these starters. They are often provided to the cheese maker in frozen, concentrated form and used to directly inoculate the cheese milk. Direct inoculation is possible because 10 to 30 times less starter is required in the manufacture of Gruyere and Emmentaler type cheeses than, for example, Cheddar cheese.

It is now well established that there are bacteriophages for both thermophilic rod and coccus bacteria.[96-101] Because of this, cheese production problems can occur, especially when pasteurized milk or other heated starter media are used in which to prepare the starter in the cheese plant. Under these conditions, cheesemakers must rely exclusively on the added starter to achieve the desired acid and proportion of rods and cocci. Infection with disturbing bacteriophages can interrupt the cheesemaking process and also adversely affect the texture and flavor characteristics of the cheese being made. Several commercial starter media are available in the U.S. for use in preparing bulk starter for use in Swiss and Italian cheese manufacture. However, there are no published data that show whether or not these media are phage inhibitory.

Another area of needed research for rod-coccus cultures is the influence that the ratio of rods and cocci have on the quality of the final cheese. In the case of Mozzarella cheese, it is felt that an excessive population of the rods can lead to a soft-bodied cheese because of the greater proteolytic ability of the rods. On the other hand, an excessive number of cocci may produce a cheese with poor meltdown characteristics when used on pizza because of the weaker proteolytic ability of the cocci. Ratios of 2:1 to 3:1 cocci to rods is considered ideal for Mozzarella cheese. Microscopic examination of fresh bulk culture smears in the cheese plant can give some indication of the relative

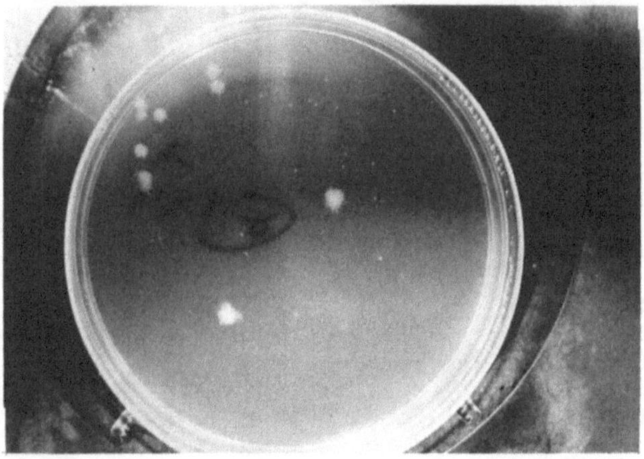

FIGURE 6. Agar plate of Driessen's LB agar[102] showing large colonies of *Lactobacillus bulgaricus* and small colonies of *Streptococcus thermophilus*.

proportion of the two types of organisms present. However, plating methods, because they measure only viable cells, are more accurate. The LB medium of Driessen et al.,[102] designed to enumerate *L. bulgaricus*, will differentiate rod and coccus types in a single plating; the rods will grow to large colonies while the cocci will be tiny colonies (see Figure 6).

Radke-Mitchell and Sandine[103] have recently reviewed some of the factors that are important in associative growth of rods and cocci in thermophilic starters. Synergism between strains is much more important in associative growth balance than temperature of incubation.[104] That is, one can combine strains of *S. thermophilus* with a low optimum growth temperature (around 35°C) with strains of *L. bulgaricus* or *L. helveticus* that have optimum temperatures around 45°C. One would expect that incubation of such cultures at 35°C would result in a predominance of rods. However, they have found that this is not the case. While pure cultures of the rod will not grow very well at 35°C and pure cultures of the coccus will not grow very well at 45°C, they both will grow reasonably well together at either temperature and yield 2:1 to 3:1 coccus to rod ratios in a mature culture. Thus, associative growth will overcome expected adverse effects[105] of diverse optimum temperatures of strains grown together.

It is clear that much less is known about thermophilic lactic acid starter cultures than is known about mesophilic starters. More research needs to be given to thermophilic starters, especially to learn more about the bacteriophages that infect them. There also is a need to develop improved starter media that are phage inhibitory for both rod and coccus phages and to study the possibility of using phage-resistant mutants in the manufacture of Italian and Swiss cheese, such as is being successfully employed for Cheddar cheese, cottage cheese, and cultured buttermilk.[22,46,48,49]

REFERENCES

1. Sandine, W. E., Starter systems for cheesemaking, *Cult. Dairy Prod. J.*, 10, 6, 1975.
2. Lawrence, R. C. and Pearce, L. E., Cheese starters under control, *Dairy Ind.*, 37, 73, 1972.
3. Lundstedt, E., Some reflections on the development of starters for the cultured dairy products industry, *Cult. Dairy Prod. J.*, 18, 10, 1983.
4. Garvie, E. I. and Farrow, J. A., *Streptococcus lactis* subsp. *cremoris* (Orla-Jensen) comb. nov. and *Streptococcus lactis* subsp. *diacetylactis* (Matuszewski et al.) nom. rev., comb. nov., *Int. J. Syst. Bacteriol.*, 32, 453, 1982.
5. Sandine, W. E., Lactic starter culture technology, in *Pfizer Cheese Monographs*, Vol. 6, Chas. Pfizer & Co., New York, 1979.
6. Sandine, W. E., New techniques in handling lactic cultures to enhance their performance, *J. Dairy Sci.*, 60, 822, 1977.
7. Hull, R. R., Control of bacteriophages in cheese factories, *Aust. J. Dairy Technol.*, 32, 65, 1977.
8. Hull, R. R., A factory derived starter selection program for bacteriophage control where single strain starters are used, in Prod. 3rd Biennial Cheese Ind. Conf., Utah State University, Logan, 1978.
9. Thomas, T. D., Heap, H. A., and Lawrence, R. C., Addition of whey to a multiple strain starter culture for casein making, *N. Z. J. Dairy Sci. Technol.*, 12, 1, 1977.
10. Thomas, T. D. and Lowrie, R. J., Starters and bacteriophage in lactic acid casein manufacture, *J. Milk Food Technol.*, 38, 275, 1975.
11. Galesloot, T. E., Hassing, F., and Stadhouders, J., Differences in phage sensitivity of starters propagated in practice and in a dairy research laboratory, Proc. 17th Int. Dairy Congr., D-491, Munich, 1966.
12. Nichols, A. A. and Ineson, P. J., Cheese starter "recovery" after attack by bacteriophage (strain dominance in multiple-strain starters), *J. Dairy Res.*, 15, 99, 1947.
13. LaGrange, W. S. and Reinbold, G. W., Starter costs in Iowa Cheddar cheese plants, *J. Dairy Sci.*, 51, 1985, 1968.
14. Collins, E. B., Action of bacteriophage on mixed strain cultures. V. Similarities among strains of lactic streptococci in commercially used cultures and use of a whey activity test for culture selection and rotation, *Appl. Microbiol.*, 3, 145, 1955.
15. Heap, H. A. and Lawrence, R. C., The contribution of starter strains to the level of phage infection in a commerical cheese factory, *N. Z. J. Dairy Sci. Technol.*, 12, 213, 1977.
16. Terzaghi, B. E., Morphologies and host sensitivities of lactic streptococcal phages from cheese factories, *N. Z. J. Dairy Sci. Technol.*, 11, 155, 1976.
17. Lawrence, R. C., Action of bacteriophages of lactic acid bacteria: consequences and protection, *N. Z. J. Dairy Sci. Technol.*, 13, 129, 1978.
18. Park, C. and McKay, L. L., Induction of prophage in lactic streptococci isolated from commerical dairy starter cultures, *J. Milk Food Technol.*, 38, 594, 1975.
19. Terzaghi, B. E. and Sandine, W. E., Bacteriophage production following exposure of lactic streptococci to ultraviolet radiation, *J. Gen. Microbiol.*, 122, 184, 1981.
20. Huggins, A. R. and Sandine, W. E., Incidence and properties of temperate bacteriophages induced from lactic streptococci, *Appl. Environ. Microbiol.*, 33, 184, 1977.
21. Lowrie, R. J., Lysogenic strains of group N lactic streptococci, *Appl. Microbiol.*, 227, 210, 1974.
22. Thunell, R. K., Sandine, W. E., and Bodyfelt, F. W., Phage-insensitive multiple-strain starter approach to Cheddar cheesemaking, *J. Dairy Sci.*, 64, 2270, 1981.
23. Sanders, M. E. and Klaenhammer, T. R., Restriction and modification in group N streptococci: effect of heat on development of modified lytic bacteriophage, *Appl. Environ. Microbiol.*, 49, 500, 1980.
24. Limsowtin, G. K. Y. and Terzaghi, B. E., Characterization of bacterial isolates from a phage-carrying culture of *Streptococcus cremoris*, *N. Z. J. Dairy Sci. Technol.*, 12, 22, 1977.
25. Gasson, M. J. and Davies, F. L., Prophage-cured derivatives of *Streptococcus lactis* and *Streptococcus cremoris*, *Appl. Environ. Microbiol.*, 40, 964, 1980.
26. Limsowtin, G. K. Y., Heap, H. A., and Lawrence, R. C., Heterogeneity among strains of lactic streptococci, *N. Z. J. Dairy Sci. Technol.*, 13, 1, 1978.
27. Gilliland, S. E., Strain balance of multiple strain lactic streptococcus concentrated cultures, *J. Dairy Sci.*, 54, 1129, 1971.
28. Lee, D. A. and Collins, E. B., Influences of temperature on growth of *Streptococcus cremoris* and *Streptococcus lactis*, *J. Dairy Sci.*, 59, 405, 1976.
29. Collins, E. B., Influence of medium and temperature on end products and growth, *J. Dairy Sci.*, 69, 799, 1976.
30. Collins, E. B., Domination among strains of lactic streptococci with attention to antibiotic production, *Appl. Microbiol.*, 9, 200, 1961
31. Hurst, A., Nisin: its preservative effect and function in the growth cycle of the producer organism, in *Streptococci*, Soc. Appl. Bacteriol. Symp. Ser. No. 7, Skinner, F. A. and Quesnel, L. B., Eds., Academic Press, London, 1966.

32. Babel, F. J., Antibiosis by lactic culture bacteria, *J. Dairy Sci.*, 60, 815, 1977.
33. Geis, A., Sing, J., and Teuber, M., Potential of lactic streptococci to produce bacteriocin, *Appl. Environ. Microbiol.*, 45, 205, 1983.
34. Kozak, W., Bardowski, J., and Dobrzanski, W. T., Lactostrepcins — acid bacteriocins produced by lactic streptococci, *J. Dairy Res.*, 45, 247, 1978.
35. Davey, G. P. and Richardson, B. C., Purification and some properties of diplococcin from *Streptococcus cremoris* 346, *Appl. Environ. Microbiol.*, 41, 84, 1981.
36. Davey, G. P., Mode of action of diplococcin, a bacteriocin from *Streptococcus cremoris* 346, *N. Z. J. Dairy Sci. Technol.*, 16, 187, 1981.
37. Collins, E. B., Action of bacteriophage on mixed strain starter cultures. I. Nature and characteristics of the nascent phenomenon, *J. Dairy Sci.*, 33, 371, 1952.
38. Tourville, D. R. and Johnstone, D. B., Lactic streptococcal phage-associated lysin. I. Lysin of heterologous streptococci by a phage-induced lysin, *J. Dairy Sci.*, 49, 158, 1966.
39. Oram, J. D. and Reiter, B., Phage-associated lysins affecting group N and group D streptococci, *J. Gen. Microbiol.*, 49, 57, 1965.
40. Reiter, B. and Oram, J. D., Group N streptococcal phage-lysin, *J. Gen. Microbiol.*, 32, 29, 1963.
41. Dahiya, R. S. and Speck, M. L., Symbiosis among lactic streptococci, *J. Dairy Sci.*, 45, 607, 1962.
42. Dahiya, R. S. and Speck, M. L., Identification of stimulatory factor involved in symbiotic growth of *Streptococcus lactis* and *Streptococcus cremoris*, *J. Bacteriol.*, 85, 585, 1963.
43. Kothari, S. L., Nambudripad, V. K. N., and Laxminarayana, H., Studies on the associative growth of some streptococci, *Milchwissenschaft*, 26, 415, 1970.
44. Lawrence, R. C., Miles-Marschall International Award Lecture, presented at the Annual Meeting of the American Dairy Science Association, Madison, Wisc., June 27, 1983.
45. Limsowtin, G. K. Y., Heap, H. A., and Lawrence, R. C., A multiple starter concept for cheesemaking, *N. Z. J. Dairy Sci. Technol.*, 12, 101, 1977.
46. Thunell, R. K., Commercial application of a defined strain starter system for cheesemaking, in Proc. 5th Biennial Cheese Ind. Conf., Utah State University, Logan, 1982.
47. Sandine, W. E., Cloning and growing virus-resistant bacteria for cheesemaking, *Proc. Agric. Conf Days*, Oregon State University, Corvallis, 1983, 81.
48. Thunell, R. K., Bodyfelt, F. W., and Sandine, W. E., Economic comparisons of Cheddar cheese manufactured with defined-strain and commercial mixed-strain cultures, *J. Dairy Sci.*, 67, 1061, 1984.
49. Thunell, R. K., Sandine, W. E., and Bodyfelt, F. W., Defined strains and phage-insensitive mutants for commerical manufacture of cottage cheese and cultured buttermilk, *J. Dairy Sci.*, 67, 1175, 1984.
50. Gilles, A. J. and Curtis, C. I., A comparison between single strain and mixed strain cultures in Cheddar cheesemaking, *Aust. J. Dairy Technol.*, 18, 66, 1963.
51. Lawrence, R. C., Heap, H. A., Limsowtin, G., and Jarvis, A. W., Cheddar cheese starters: current knowledge and practices of phage characteristics and strain selection, *J. Dairy Sci.*, 61, 1181, 1978.
52. Horrall, B. E. and Elliker, P. R., An activity test for Cheddar and Cottage cheese starters, *J. Dairy Sci.*, 33, 245, 1950.
53. Pearce, L. E., Activity tests for cheese starter cultures, *N. Z. J. Dairy Technol.*, 4, 246, 1969.
54. Heap, H. A. and Lawrence, R. C., Recent modifications to the New Zealand activity test for Cheddar cheese starters, *N. Z. J. Dairy Sci. Technol.*, 15, 91, 1981.
55. Richardson, G. H., Ernstrom, C. A., and Hong, G. L., The use of defined starter strains and culture neutralization in United States cheese plants, *Cult. Dairy Prod. J.*, 16, 11, 1981.
56. Heap, H. A. and Lawrence, R. C., The selection of starter strains for cheesemaking, *N. Z. J. Dairy Sci. Technol.*, 11, 16, 1976.
57. Henning, D. R., Black, C. H., Sandine, W. E., and Elliker, P. R., Host-range studies of lactic streptococcal bacteriophages, *J. Dairy Sci.*, 51, 1, 1968.
58. Gamay, A. Y., Richardson, G. H., and Brown, R. J., Simplified pairing of lactic streptococcal strains for cheese manufacture, *J. Dairy Sci.*, 67, 1181, 1984.
59. Sullivan, J. J., Schlipalius, L. E., Mou, L., and Jago, G. R., Heat lability of starter cultures and its use in selecting potentially non-bitter strains, *Aust. J. Dairy Technol.*, 29, 71, 1976.
60. Lawrence, R. C., Creamer, L. K., Gilles, J., and Martley, F. G., Cheddar cheese flavour. I. The role of starters and rennets, *N. Z. J. Dairy Sci. Technol.*, 7, 32, 1972.
61. Martley, F. G. and Lawrence, R. C., Cheddar cheese flavour. II. Characteristics of single strain starters associated with good or poor flavour development, *N. Z. J. Dairy Sci. Technol.*, 7, 38, 1972.
62. Lowrie, R. J., Lawrence, R. C., and Pearce, L. E., Cheddar cheese flavour. III. The growth of lactic streptococci during cheesemaking and the effect on bitterness development, *N. Z. J. Dairy Sci. Technol.*, 7, 44, 1972.
63. Lowrie, R. J. and Lawrence, R. C., Cheddar cheese flavour. IV. A new hypothesis to account for the development of bitterness, *N. Z. J. Dairy Sci. Technol.*, 7, 51, 1972.

64. Breheny, S., Kanasaki, M., Hillier, A. J., and Jago, G. R., Effect of temperature on the growth and acid production of lactic acid bacteria. II. The uncoupling of acid production from growth, *Aust. J. Dairy Technol.*, 30, 145, 1975.
65. Hillier, A. J., Kanaski, M., and Jago, G. R., Effect of temperature on the growth and acid production of lactic acid bacteria. III. The influence of added growth supplement, *Aust. J. Dairy Technol.*, 30, 149, 1975.
66. Whitehead, H. R. and Cox, G. A., Bacteriophage phenomena in cultures of lactic streptococci, *J. Dairy Res.*, 7, 55, 1936.
67. Czulak, J., Bant, D. J., Blyth, S. C., and Grace, J., A new cheese starter system, *Dairy Ind. Int.*, 44, 17, 1979.
68. Jarvis, A. W., The use of whey-derived phage-resistant starter strains in New Zealand cheese plants, *N. Z. J. Dairy Sci. Technol.*, 15, 25, 1981.
69. Limsowtin, G. K. Y. and Terzaghi, B. E., Phage resistant mutants: their selection and use in cheese factories, *N. Z. J. Dairy Sci. Technol.*, 11, 251, 1976.
70. Marshall, R. J. and Berridge, N. J., Selection and some properties of phage-resistant starters for cheesemaking, *J. Dairy Res.*, 43, 449, 1976.
71. Huggins, A. R. and Sandine, W. E., Differentiation of fast and slow milk-coagulating isolates in strains of lactic streptococci, *J. Dairy Sci.*, accepted for publication, 1984.
72. Richardson, G. H., Proteinase negative cultures for cheesemaking, in Proc. 5th Biennial Cheese Ind. Conf., Utah State University, Logan, 1982.
73. Pearce, L. E., Skipper, N. A., and Jarvis, B. D. W., Proteinase activity in slow lactic acid-producing variants of *Streptococcus lactis*, *Appl. Microbiol.*, 27, 933, 1974.
74. Elliker, P. R., Anderson, A. W., and Hannesson, G., An agar culture medium for lactic acid streptococci and lactobacilli, *J. Dairy Sci.*, 39, 1611, 1956.
75. Mills, O. E. and Thomas, T. D., Bitterness development in Cheddar cheese: effect of the level of starter proteinase, *N. Z. J. Dairy Sci. Technol.*, 15, 131, 1980.
76. Richardson, G. H., Ernstrom, C. A., Kim, J. M., and Daly, C., Proteinase negative variants of *Streptococcus cremoris* for cheese starters, *J. Dairy Sci.*, 66, 2278, 1983.
77. Shelaih, M. S., Gamay, A. Y. E., Wright, S. L., and Richardson, G. H., Temperature sensitivities of proteinase negative varients of lactic streptococci, *J. Dairy Sci.*, 66, 2287, 1983.
78. Pearce, L. E., A survey of bulk starter preparation and handling in New Zealand cheese factories, *N. Z. J. Dairy Sci. Technol.*, 8, 17, 1973.
79. Harvey, R. J., Damage to *Streptococcus lactis* resulting from growth at low pH, *J. Bacteriol.*, 90, 1330, 1965.
80. Richardson, G. H., Cheng, C. T., and Young, R., Lactic bulk culture system utilizing a whey-based bacteriophage inhibitory medium and pH control. I. Applicability to American style cheese, *J. Dairy Sci.*, 60, 378, 1977.
81. Richardson, G. H. and Pearce, L. E., A laboratory batch fermenter with differential electrode pH control, *N. Z. J. Dairy Sci. Technol.*, 11, 83, 1976.
82. Ausavanodom, N., White, R. S., Young, G., and Richardson, G. H., Lactic bulk culture system utilizing whey-based bacteriophage inhibitory medium and pH control. II. Reduction of phosphate requirements under pH control, *J. Dairy Sci.*, 60, 1245, 1977.
83. Chen, Y. and Richardson, G. H., Lactic bulk culture system utilizing whey-based bacteriophage inhibitory medium and pH control. III. Applicability to cottage cheese manufacture, *J. Dairy Sci.*, 60, 1252, 1977.
84. Reddy, K. P. and Richardson, G. H., Lactic bulk culture system utilizing whey-based bacteriophage inhibitory medium and pH control. IV. Applicability to Italian and Swiss cheese cultures, *J. Dairy Sci.*, 60, 1527, 1977.
85. Limsowtin, G. K. Y., Heap, H. A., and Lawrence, R. C., A new approach to the preparation of bulk starter in commercial cheese plants, *N. Z. J. Dairy Sci. Technol.*, 15, 219, 1980.
86. Jonas, L. L., Ernstrom, C. A., and Richardson, G. H., Major savings realized in the production of bulk lactic culture with whey-based phage inhibitory medium provided with pH control, *Cult. Dairy Prod. J.*, 12, 12, 1977.
87. Willrett, D. L., Sandine, W. E., and Ayres, J. W., Evaluation of a new pH-controlled bulk starter medium, *Cheese Reporter*, 103, 8, 1979.
88. Morris, C. E., "Breakthrough" in starter medium, *Food Eng.*, 52, 38, 1980.
89. Sandine, W. E. and Ayres, J. W., U.S. Patent 4,282,255, 1981.
90. Kosikowski, F., *Cheese and Fermented Milk Foods*, 2nd ed., Edwards Brothers, Ann Arbor, Mich., 1977.
91. Reinbold, G. W., Italian cheese varieties, in *Pfizer Cheese Monographs*, Vol. 1, Chas. Pfizer & Co., New York, 1963.
92. Reinbold, G. W., Swiss cheese varieties, in *Pfizer Cheese Monographs*, Vol. 5, Chas. Pfizer & Co., New York, 1972.

93. Wood, N. J., Culture and starter programs used in cheese plants in Italy, *Dairy Ice Cream Field*, 160, 64C, 1977.
94. Accolas, J. P. and Auclair, J., Thermophilic lactic starters, *Ir. J. Food Sci. Technol.*, 7, 27, 1983.
95. Auclair, J. and Accolas, J. P., Use of thermophilic lactic starters in the dairy industry, *Antonie van Leeuwenhoek; J. Microbiol. Serol.*, 49, 313, 1983.
96. DeKlerk, H. C., Coetzee, J. N., and Fourie, J. T., The fine structure of *Lactobacillus* bacteriophages, *J. Gen. Microbiol.*, 38, 35, 1965.
97. Sozzi, T., Maret, R., and Poulin, J. M., Study of plating efficiency of bacteriophages of thermophilic lactic acid bacteria on different media, *Appl. Environ. Microbiol.*, 32, 131, 1976.
98. Yokokur, T., Kodaira, S., Ishiwa, H., and Sakurai, T., Lysogeny in lactobacilli, *J. Gen. Microbiol.*, 84, 277, 1974.
99. Accolas, J. P. and Spreeman, H., The morphology of six bacteriophages of *Streptococcus thermophilus*, *J. Appl. Bacteriol.*, 47, 135, 1979.
100. Accolas, J. P. and Spillman, H., Morphology of bacteriophages of *Lactobacillus bulgaricus*, *L. lactis*, and *L. helveticus*, *J. Appl. Bacteriol.*, 47, 309, 1979.
101. Reinbold, G. W., Reddy, M. S., and Hammond, E. G., Ultrastructure of bacteriophages active against *Streptococcus thermophilus*, *Lactobacillus bulgaricus*, *Lactobacillus lactis*, and *Lactobacillus helveticus*, *J. Food Protect.*, 45, 119, 1982.
102. Driessen, R. M., Ubbels, J., and Stadhouders, J., Continuous manufacture of yogurt. I. Optimal conditions and kinetics of the prefermentation process, *Biotechnol. Bioeng.*, 19, 821, 1977.
103. Radke-Mitchell, L. and Sandine, W. E., Associative growth and differential enumeration of *Streptococcus thermophilus* and *Lactobacillus bulgaricus*. A review, *J. Food Protect.*, 47, 245, 1984.
104. Radke-Mitchell, L., Influence of Temperature on Associative Growth of *Streptococcus thermophilus* and *Lactobacillus bulgaricus*, M.S. thesis, Oregon State University, Corvallis, 1983.
105. Martley, F., Temperature sensitivities of thermophilic starter strains, *N. Z. J. Dairy Sci. Technol.*, 18, 191, 1983.

Chapter 11

CONCENTRATED STARTER CULTURES

Stanley E. Gilliland

TABLE OF CONTENTS

I. IMPORTANCE TO THE INDUSTRY

Concentrated starter cultures are defined as starter cultures that are grown under closely controlled conditions, concentrated into a small volume, and frozen or dried for storage and transportation. These have provided improved starter cultures that are easier to utilize in the manufacture of cultured foods than traditional forms of starters. In the traditional or the conventional approach, stock cultures were maintained in processing plants via regular subcultures in the plant laboratory. It was necessary to build up the volume of the culture through successive stages or subcultures to have sufficient volume (bulk starter culture) to inoculate the product to be cultured (Figure 1). The newer approach involving the use of frozen or dried concentrated cultures has made it possible to inoculate either the bulk starter vat or the product itself directly.

There are several advantages for using the concentrated starter cultures:[1,2] (1) the activity of the cultures can be carefully monitored and controlled prior to shipping to the processing plant, (2) the maintainance of stock cultures in the processing plant can be eliminated, (3) less labor is required, and (4) starter culture rotation systems are easier to establish in order to help maintain control over bacteriophage. Along with the advantages for using these types of starter cultures are several disadvantages, including the requirement of adequate storage space at a proper low temperature both for shipping and maintaining the cultures so that they do not lose activity. Furthermore, not all starter cultures that have been useful over the years for producing high quality cultured food products are suitable for preparing concentrated starter cultures. This can, of course, be overcome by proper selection of cultures for preparing the concentrates.

II. ACCEPTABILITY IN RELATION TO QUALITY OF CULTURED FOODS

Many traditional cultured food products have been successfully manufactured using concentrated cultures.[2-9] Because of the control that can be exerted over the production of concentrated starter cultures to ensure optimal activity, we can expect greater consistency with regard to the production of high quality cultured foods. Cultured buttermilk of consistent high quality can be manufactured using frozen concentrated starter cultures.[2] In the manufacture of Cheddar cheese the yield has been reported to be slightly higher when frozen concentrated starter cultures are used than when conventional cultures are used.[7] No significant differences were observed in flavor and texture of the cheeses. The difference in yield may be related to the fact that the use of conventional or traditional starter cultures would have involved inoculating the cheese vat with a considerable volume of bulk starter. Some of the milk proteins denatured in heating of the milk used for preparation of the bulk starters may have been lost as fines in the whey during the cheese manufacturing process, thus resulting in an apparent lower yield than when the concentrated cultures were used. Such denatured proteins would not have been added to the cheese vat when inoculated with the concentrated cultures.

Frozen concentrated starter cultures also have been very important in the production of cultured meat products. The inoculation of meat for preparing certain fermented sausage products with a massive number of bacteria contained in a concentrated starter culture makes it possible to have greater control over the fermentation process in producing these products.

In addition to improving procedures for making existing products, the development of frozen concentrated starter cultures has made it possible to produce new products. A fine flavored cottage cheese can be produced by adding a creaming mixture that had

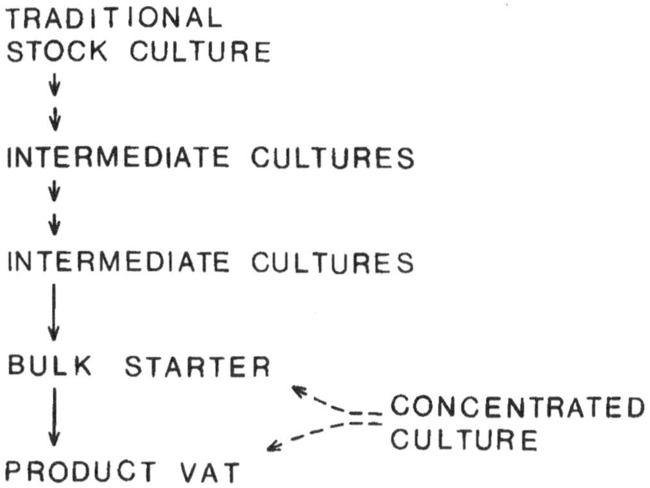

FIGURE 1. Steps involved in use of traditional starter cultures and concentrated starter cultures.

been cultured with *Leuconostoc cremoris* (formerly *L. citrovorum*). The improved flavor of the product is due in part to the diacetyl produced during the fermentation of citrate in the cream by the leuconostocs. One of the problems associated with this particular procedure has been the time involved in growing the leuconostoc organisms in the creaming mixture. However, the development of the frozen concentrated cultures of *L. cremoris* has made it possible to add the bacteria to an acidified creaming mixture followed by a brief incubation period for the production of diacetyl and other volatile compounds.[10] The leuconostoc culture, which is relatively slow growing organism, is not required to grow in the creaming mixture and thus much time is saved in the process. The volatile components produced by the cells from the concentrated cultures under such conditions are similar to those produced by conventional milk cultures.[11]

Another new product made possible by the development of frozen concentrated starter cultures is nonfermented acidophilus milk. Formerly, *Lactobacillus acidophilus* was available only in a fermented milk product that had a much less than desirable taste. The new product, which first became commercially available in 1975, has the same flavor as pasteurized milk yet contains approximately 2 million cells of *L. acidophilus* per milliliter. Compared to the original fermented product, the nonfermented acidophilus milk provides the consumer with a more desirable source of this organism which has been shown to be beneficial in the intestinal tract.[12-14]

III. GENERAL PROCEDURES FOR PRODUCING CONCENTRATED CULTURES

Methods for preparing concentrated starter cultures include several variations from those used for the traditional or conventional starter cultures.[15] Cells of the starter bacteria are grown under strictly controlled conditions in a medium from which the cells can be easily harvested or concentrated into a small volume. The objective in producing the cells is to produce the maximum number per unit volume of growth medium that will contain the necessary compliment of enzymes to perform properly in manufacturing cultured food products. The general steps involved in the preparation of concentrated cultures include: (1) inoculation of the suitable growth medium, (2)

INOCULATE GROWTH MEDIUM

↓

INCUBATE

↓

HARVEST CELLS

↓

ADD SUSPENDING MENSTRUM

↓

FREEZE (OR DRY)

↓

STORE

FIGURE 2. Generalized steps in preparing concentrated starter cultures.

incubation until the culture reaches maximum population, (3) concentration of the cells into the small volume of a desirable suspending menstrum, (4) freezing (or drying), (5) storing, (6) shipping to the processing plant, and (7) using the concentrated culture in preparing the cultured product (Figure 2). It is necessary that the bacteria in the concentrated starter culture maintain activity and/or desirable characteristics throughout the production procedures, storage, transportation, and subsequent storage at the processing plant prior to use.

IV. PRODUCTION OF CELL CROPS

A. Growth Medium

One of the first factors to be considered in producing cell crops of the starter cultures is the growth medium. It should provide all the necessary nutrients for maximum growth of the organism as well as ensuring that the resulting cells have the necessary enzymes to produce the desired changes in the food to be cultured. Furthermore, the resulting cultures should possess optimum activity and be stable to storage for extended periods of time. The choice of the growth medium may be controlled by such factors as costs, ability to produce high numbers of bacterial cells containing suitable activity, and the ease with which cells can be harvested.[1] Quite a variety of different media have been reported for the production of cells of dairy starter cultures. These have involved various concentrations of peptone, yeast extract, and sugar (primarily lactose).[16-18] Media including these types of ingredients have been chosen primarily for research studies because cells can be harvested easily and because control can be exercised over the concentration of nutrients in the growth medium relatively easily. The concentration of the various ingredients can be easily altered to determine the composition favoring maximum growth of the culture and subsequent performance of the resulting concentrated starter culture.

The use of cheese whey as a base medium for producing cell crops of starter culture has been attractive primarily because it is essentially a waste product and relatively inexpensive.[19] Furthermore, it contains nutrients that many of the starter culture bacteria can readily utilize for growth. The use of cheese whey alone has not been too successful because of the limitations for some nutrients that support maximum growth of starter culture bacteria. Much of the research using cheese whey has involved supplementing the whey. Some have used proteolytic enzymes to hydrolyze or partially hydrolyze the whey proteins to improve it as a medium for growing the bacteria. Even with such enzymatic treatment, whey still has not been considered adequate for maximum growth of starter culture bacteria.[20,21] In addition, if not handled properly during preparation, the medium develops a precipitate during the heat sterilization process. The rate and intensity of heat treatment appears to be important in controlling the development of this precipitated matter.[20] If precipitation occurs, clarification steps might be necessary in order to remove it prior to inoculating the medium with the starter culture so that the precipitated material does not appear in the concentrated starter culture.[21]

Skim milk (or reconstituted nonfat milk solids) has been the choice medium of a number of investigators for producing cells of the lactic streptococci for concentrate preparation.[22-26] The primary advantage of using this medium over the whey-based media or the peptone-based media is that it represents the same composition of the medium that will be used in the manufacture of the cultured milk products.[23-25] For dairy starter cultures, it is generally accepted that at least some milk solids should be included in the growth medium. This is not only to help assure that the concentrated culture will have the necessary enzymes to perform well when inoculated into the milk vat, but also in order to maintain the balance among strains in multiple strain starter cultures of the lactic streptococci.[22,23]

The preparation of concentrated starter cultures from cells grown in a skim milk medium may be inefficient since it is difficult to harvest the cells from milk by centrifugation. This can, however, be overcome by adding sterile sodium citrate (final concentration 1 to 2%) to the culture that had been maintained at a desirable pH above that of the isoelectric point of casein.[23,24,26] The sodium citrate is added to the culture after maximum population has been achieved or at the point where harvesting the bacterial cells is desired. The sodium citrate solubilizes the milk proteins and thus clarifies the milk so that the bacteria can be easily removed by centrifugation. In some cases, especially with the lactic streptococci, attempts to use sodium citrate to clarify the milk medium have not been very successful (unpublished observations). Thus, not all cultures may be suitable to use in a system employing skim milk as a growth medium for producing cells for concentrated starter cultures.

It may be necessary to supplement the growth medium with materials other than nutrients needed for growth of the starter culture bacteria in order to ensure that they will be able to carry out the desired function in preparing a cultured product. For instance, for concentrated cultures of *L. cremoris* to produce diacetyl and other flavor components in an acidified milk (such as mentioned previously in preparing a creaming mixture for cottage cheese) requires that the concentrate be prepared from cells grown in a medium supplemented with sodium citrate.[10,11] If citrate is omitted from the growth medium the resulting concentrated starter culture will not be able to produce diacetyl from citrate. Apparently, the cells grown without sodium citrate in the medium lack the enzyme citrate permease and thus cannot utilize citrate.

B. Growth Conditions

In selecting the temperature at which to grow cell crops of starter cultures the optimum growth temperature for the organisms as well as the temperature at which the resulting concentrated starter culture will be used must be considered. Usually, the optimum growth temperature of the organism should be used unless an advantage could be obtained with regard to maximum cell production without adversely affecting the performance of the resulting concentrate by using some other growth temperature.

In addition to providing the necessary nutrients and a favorable temperature for the bacteria, it is generally possible to greatly increase the number of cells of starter cultures by maintaining the pH of the growth medium at a level favorable for their growth. Low pH resulting from the production of lactic acid by these bacteria limits the total amount of growth that can be achieved. Maintenance of the pH of the growth medium in a favorable range can result in greater than tenfold increases in a number of cells of lactic streptococci produced per unit volume of growth medium as compared to a culture grown without pH control.[16,25,27] The method of controlling the pH at a desirable level is important. Proper maintenance of the desired pH of the culture during growth can best be achieved by the use of electronic pH control equipment designed to continuously monitor the pH of the growth medium and add neutralizer automatically as needed to maintain the desired pH. For best results using pH control, both the desired pH range and the alkali used to maintain the desired pH must be considered. Most work reported concerning the evaluation of neutralizers has been done using lactic streptococci. The optimum pH for these organisms with regard to the production of maximum cell numbers is in the range of 6 to 6.5.[4,16,19,25,28,29] Some of the variations in the reported pH optima for this group of organisms may have been due to variations that existed with regard to other growth conditions such as composition of the growth media. Sodium hydroxide, as a neutralizer, is less desirable than ammonium hydroxide or calcium hydroxide.[4,16,17,25,29] The use of ammonium hydroxide and/or calcium hydroxide to maintain the pH of cultures of lactic streptococci during growth results in higher cell yields than when sodium hydroxide is used, however, the resulting cells appear to be less active.[4,16] Concentrated cultures of lactic streptococci prepared from cells grown at pH 6 in which ammonium hydroxide was used as the neutralizer, however, regained activity during the first subculture in milk. The decreased activity of the organisms in the concentrated culture was related to the decrease in proteinase activity of the lactic streptococci. The proteinase activity was, at least in part, regained during the first subculture in milk. This points out an advantage for using the concentrated starter cultures to inoculate the bulk culture rather than the final product vat. The bulk starter step allows the bacteria one subculture in which to regain any activity that may have been lost or reduced as a result of being grown at controlled pH.

The growth of mixed species starter cultures (*S. lactis, S. cremoris, L. cremoris,* or *S. diacetilactis*) at pH 6 or 6.5 using sodium hydroxide as the neutralizer produces a negative effect on the balance among species compared to that observed in traditional milk cultures.[25] The use of sodium hydroxide as a neutralizer favors the aroma bacteria. The use of ammonium hydroxide as a neutralizer does not have this deleterious effect on the balance among species. Another negative effect resulting from the use of sodium hydroxide as a neutralizer is that resulting concentrated cultures do not survive as well when frozen at −30°C as do concentrated cultures prepared from cells grown with ammonium hydroxide as a neutralizer.[17,29]

Another neutralizer that should be equally as good as ammonium hydroxide is anhydrous ammonia gas. An advantage to using the anhydrous ammonia would be that introduction of the gaseous ammonium into the starter medium for maintaining the pH would not introduce additional liquid, thus diluting the culture. Little or no work has been reported concerning the comparison of other alkaline materials for neutralizers. A mixture of ammonium hydroxide and calcium carbonate has been shown to be effective for producing cells of *L. acidophilus.*[20] At pH 6 this neutralizer provides slow release of carbon dioxide which appears to benefit growth of these lactobacilli. There is a need for additional research to study the effect of different neutralizers on producing cells of all species of starter culture bacteria for concentrated cultures. Research also is needed with regard to determining the optimum pH for producing species other than lactic streptococci. This should include not only studies relating to the production of maximum cell numbers, but also to producing cells that have maximum biological activity and thus will perform well in producing cultured food products.

At present, the production of cell crops of starter bacteria is best achieved by the batch culture procedure, which is simpler and more convenient than a continuous culture system. In the batch culture system the stage in the growth curve at which the cells are harvested may in some cases be critical with regard to producing cultures having optimum activity. This factor has received limited attention. For the lactic streptococci it has been suggested that the cells should be harvested just prior to reaching the stationary phase of growth.[30] However, it also has been shown that continued incubation of these cells for periods greater than 6 hr beyond the point at which they entered the stationary phase caused no loss of activity of the resulting concentrated culture.[16] Further research is needed concerning physiological age at which cells of starter bacteria are harvested with regard to activity of the resulting concentrated starter culture as well as the stability of the concentrated culture during storage.

The problems associated with continuous cultures include possible failures due to undesirable contamination, the requirement for more complex equipment, and problems associated with production schedules in preparing concentrated starter cultures.[4,31] It would be very difficult to have sufficient culture continuously available from a continuous culture system in order to supply a high speed centrifuge used to prepare the concentrated culture. Other major problems that have been noted in the use of continuous cultures is the possible development of bacteriophage during the culturing process as well as the loss of enzyme activity for the culture.[4] In spite of these limitations, concentrated starter cultures have been made successfully from continuous cultures.[4,21,32] Continuous cultures of lactic streptococci have been maintained for up to 2 weeks in a trypsin digested skim milk medium.[21] Frozen concentrated starter cultures prepared from such cells have been used successfully in the manufacture of Cheddar cheese.[32]

C. Factors Limiting Maximum Population

To improve efficiency of producing cell crops of the starter culture bacteria, considerable attention has been focused on ways of increasing the number of cells produced per unit volume of growth medium. This has involved not only making sure that adequate nutrients are supplied to prevent nutrient limitation, but also studies of other factors that limit the growth of the bacteria. As mentioned earlier low pH due to the production of lactic acid is a major factor in controlling growth of starter culture bacteria. It is, however, fairly easy to maintain the pH of the starter culture at a desired level in order to provide a significant increase in the number of cells per unit volume. Maintenance of pH at a desirable level during the growth of the cells requires, in many cases, a fairly vigorous and continuous agitation of the growth medium. The increased oxgyen incorporated into the medium during these types of procedures can result in the production of hydrogen peroxide by the starter cultures which can cause autoinhibition of the bacteria.[33] Growth of the lactic streptococci in a liquid medium maintained at pH 6 to 6.5 results in some reduction of growth rate due to the accumulation of hydrogen peroxide.[1,34] This could be overcome by the addition of sterilized catalase to the growth medium or perhaps through the addition of other compounds such as reducing agents, pyruvate, and ferrous sulfate, all of which prevent hydrogen peroxide accumulation in lactic cultures.[33] In some cases a direct oxygen toxicity effect on the lactic streptococci has been observed that apparently does not involve the accumulation of hydrogen peroxide.[34] Another approach to preventing the influence of excess aeration is to continuously sparge an inert gas through the medium during growth of the bacteria. It also has been reported that reduced growth of some lactic streptococci in agitated cultures is due to the depletion of carbon dioxide in the growth medium.[35]

Sparging of the growth medium with carbon dioxide can be an effective means of overcoming the effect of excess aeration as well as supply the carbon dioxide that is apparently needed for optimum growth of some of the starter culture bacteria.[19,28,36] Nitrogen gas, on the other hand, has been reported to have a deleterious effect on the growth of some cultures.[28]

The accumulation of toxic metabolites other than hydrogen peroxide also can be important in controlling growth of the bacteria. Lactic streptococci grown at pH 6 accumulate D-leucine which is inhibitory to the culture.[37] The cessation of growth by lactic acid bacteria maintained at constant pH levels is probably due to a combination of factors including lactate accumulation.[26,27] In fact, the accumulation of lactate salts may be the major contributing factor. Further investigation is needed concerning the accumulation of toxic metabolites and how to remove or inactivate them in order to permit increased growth of the bacteria.

Just as nutrient limitation may influence the maximum numbers of cells produced in a growth medium, an excess of nutrients can also limit the production of bacterial cells. *L. acidophilus* grown in a pepsinized whey-based medium at pH 6 reached a maximum population after approximately 10 hr in a medium containing 2.5 or 5% whey solids and remained fairly constant for an additional 12 to 15 hr.[20] However, when grown in a medium containing 7.5% whey solids, maximum population was reached at 10 to 12 hr and then the culture entered a death phase (Figure 3). This was not observed in the static cultures of the organism in any of the concentrations of whey-based medium. Apparently, in the medium containing the 7.5% whey solids, there was an excess of some component that resulted in the organism entering the death phase. The balance among nutrients in the growth medium may be a very critical factor in controlling the production of a culture that will have the ability to survive when grown at controlled pH.

V. STORAGE STABILITY

To be useful in the cultured foods industry, concentrated starter cultures must survive storage for extended periods. The cultures must not only maintain viability, but activity and/or other desirable characteristics. Producing concentrated cultures that are stable during storage represents perhaps the biggest challenge.

A. Frozen Storage

Early reports on frozen concentrated cultures of lactic streptococci prepared from cells grown in a tryptone-yeast extract medium indicated that there was little loss of viability or activity during 9 months of storage at −20°C.[38] These concentrated cultures contained from 2.5×10^{10} to 5.5×10^{10} cells per milliliter. At this storage temperature the cells survived much better when frozen at pH 7 than when frozen at pH 5. Thus, the pH of the culture at freezing is a very important factor in governing how well the culture will survive during frozen storage. While for some strains of lactic streptococci, storage at −20 to −30°C is satisfactory for maintaining the culture, there is variation among strains in that some will not survive well when frozen at these temperatures.[3,30,38-40] Most of the cultures can be stored for short times at −20 to −40°C, however, lower temperatures should be used for long-term storage. The most deleterious temperatures for storage of the lactic streptococci in general appear to be −20°C or above. The best temperature for storage, especially long term, is −196°C (in liquid nitrogen). Lloyd and Pont[4] reported that the survival of lactic streptococci in concentrated cultures was 42% at −20°C, 93 to 99% at −40°C, and 100% at −110°C (vapor phase of liquid nitrogen). Concentrated cultures of lactic streptococci remain viable and lose no activity for periods up to 231 days at −196°C.[16]

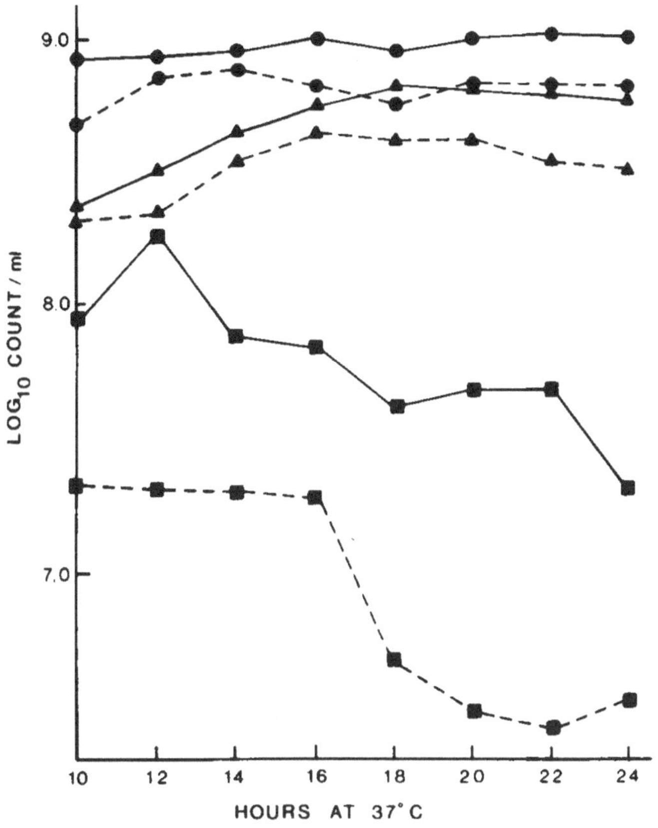

FIGURE 3. Growth of *Lactobacillus acidophilus* NCFM at pH 6.0 in pepsinized whey based media. (●) 2.5% Whey solids (averages from 6 trials); (▲) 5.0% whey solids (averages from 6 trials); (■) 7.5% whey solids (averages from 4 trials); solid lines = counts on MRS agar; broken lines = counts on MRSO agar. (From Mitchell, S. L. and Gilliland, S. E., *J. Dairy Sci.*, 66, 712, 1983. With permission.)

Lactic streptococci grown in a medium containing tryptone, yeast extract, and lactose at pH 6 survived better at −17°C than did cultures of the same strains grown in the same medium without pH control.[39] There was no clear explanation for this difference in storage stability. The ability of these cultures to survive freezing at −17°C was related to the capsular content for cells of strains grown with and without pH control during growth. As the cellular glucose content increased so did the survival of the resulting concentrated culture during frozen storage. There was, however, no indication that cells grown at pH 6 contained more capsular material than those grown without pH control. In other studies, lactic streptococci grown at pH 6 or 6.5 using ammonium hydroxide as a neutralizer survived better than did the same cultures grown at pH 6 or 6.5 with sodium hydroxide as the neutralizer.[17,29]

The freezing and storage of concentrated cultures of the lactic streptococci in liquid nitrogen at −196°C provides a very satisfactory means of storage regardless of whether the cells have been grown in a peptone-yeast extract broth or in a milk based medium.[40] However, freezing of concentrated suspensions of *L. bulgaricus* at −196°C can result in tremendous loss of viability unless the cells have been grown in an appropriate medium.[41] Cells of the lactobacilli grown in a medium supplemented with Tween® 80 survived freezing and storage at −196°C, whereas those grown without Tween® 80 did not. Similar results have also been reported for cultures frozen at −17°C.[42] Oleic acid contained in the Tween® 80 was identified as a component responsible for this increase in survival of the lactobacilli during frozen storage.[43] The inclusion of oleic acid in the growth medium resulted in the cells having increased levels of a C_{19} cyclopropane fatty acid in the lipid fraction. The amount of this C_{19} fatty acid was significantly related to the ability of the cultures to survive. As the cellular content of this particular fatty acid increased, so did the survival level of the culture during frozen storage (Chapter 4, Figure 5). Since the majority of fatty acids in bacterial cells are located in the membrane of the cells, it may be assumed that the membrane composition is very important in determining the ability of the cells to survive during freezing and frozen storage. It has been theorized that the presence of the C_{19} cyclopropane fatty acid made the membrane of the cell more flexible, thus preventing damage resulting from freezing and thawing. Cells of the lactic streptococci grown without oleic acid in the growth medium contain higher levels of the C_{19} cyclopropane fatty acid than do cells of *L. bulgaricus*.[39] This may explain why the lactic streptococci survived better at −196°C than did the lactobacilli regardless of the growth medium. However, it should be pointed out that under certain conditions there was a positive relationship between the ability of the lactic streptococci to survive at −17°C and the amount of octadecenoic acid in the lipid fraction of the cells.[39] The ratio of unsaturated to saturated fatty acids in the celluar lipids also appears in some cases to be related to the ability of streptococci and lactobacilli to survive freezing.[42] As this ratio increased so did the survival at −17°C. This further suggests that the membrane composition of the bacteria is important in enabling them to survive during frozen storage.

Over the years, many attempts have been made to add different substances (referred to as cryoprotective agents) to starter cultures in order to improve their ability to survive freezing and frozen storage. Glycerol is generally accepted as a suitable cryoprotective agent for concentrated cultures to help prevent loss of viability (or activity) during frozen storage. However, there still remains a question as to whether cryoprotective agents are necessary in all cases for frozen concentrated cultures. Even though glycerol is accepted as the most suitable, there is variation among cultures (especially of the lactic streptococci) with regard to the ability of the glycerol to provide protection during frozen storage.[24,29] Some strains are adequately protected, others are not. Stadhouders et al.,[24] in a study of optimum freezing and storage conditions, reported that 10% gylcerol provided variable results with regard to protecting lactic streptococci at −37°C. Dimethyl sulfoxide (DMSO) is an excellent cryoprotective agent, but it would not be desirable for use in food. Supplementing the concentrated cultures with 7.5% lactose prior to freezing in their studies permitted storage for up to 3 months at −37°C with no loss in activity. However, in another report, lactose was ineffective.[29] Most studies dealing with frozen concentrated cultures indicate that no cryoprotective agents are needed if the cultures have been produced under proper conditions and if they are frozen and maintained at −196°C.

B. Dried Storage

Considerably less effort has been made toward developing freeze-dried concentrated cultures than frozen concentrated starter cultures. Freeze-dried concentrated cultures have been used successfully to manufacture Cheddar cheese, however, such a culture is considered to be quite expensive.[9,23] Once properly prepared, such a culture would have the advantage of being easier to ship to the processor and store at the processing plant than the frozen concentrated cultures.

Most research on freeze-dried concentrated cultures has dealt with the lactic streptococci.[8,23,26,44,45] In general, the results have shown that the freeze-drying process has a deleterious effect on the culture resulting in death and reduced activity. The loss in activity resulting from the freeze-drying process can however be offset to a certain extent by adjusting the inoculation level used in preparing the cultured product. In addition to the freeze-drying process itself killing some of the starter culture bacteria, additional death occurs during subsequent storage of the dried culture. Some have suggested that this death occurring after drying is possibly due to oxygen damage.[45] If oxygen damage is an important factor it could perhaps be minimized by adding reducing agents to the culture and/or eliminating oxygen from the environment of the dried culture (i.e., vacuum sealing or storing the culture in an inert gas atmosphere). The addition of 5% monosodium glutamate (MSG) to cultures of lactic streptococci prior to drying has been shown to provide some protection to the culture during the drying process.[26] The addition of 7% lactose or sucrose in other situations has also been shown to provide protection to the cultures.[23]

Just as the growth conditions under which the cells are produced for concentrated cultures can influence the ability of the cultures to survive frozen storage, it might also influence the ability of the cultures to survive the freeze-drying process. This represents an area that needs further investigation. Factors to consider include not only the composition of the growth medium with regard to nutrients and/or supplements, but also to the conditions such as pH and temperature under which the cells are grown. Undoubtedly, the factors influencing the ability of the cultures to survive the freeze-drying process must be more complex than those involved in the cultures surviving freezing. However, with additional research efforts, it should be possible to improve the ability of cultures to survive the freeze-drying process.

REFERENCES

1. Stanley, G., The manufacture of starters by batch fermentation and centrifugation to produce concentrates, *J. Soc. Dairy Technol.*, 30, 36, 1977.
2. Speck, M. L. and Gilliland, S. E., Concentrated starters and the manufacture of buttermilk, *Cult. Dairy Prod. J.*, 10, 10, 1975.
3. Accolas, J. P. and Auclair, J., Conservation a l'etat congelé de suspensions de bactéries lactiques concentrées sour faible. I. Bactéries lactiques mesaphiles. (Storage of highly concentrated suspension of lactic acid bacteria in the frozen state. I. Mesophilic lactic acid bacteria), *Lait*, 47, 253, 1967.
4. Lloyd, G. T. and Pont, E. G., Some properties of frozen concentrated starters produced by continous culture, *J. Dairy Res.*, 40, 157, 1973.
5. Rousseaux, P., Vassal, L., Valles, E., and Auclair, J., Utilisation en fromagerie de Gruyère de suspensions concentrées et congelées de bactéries lactiques thermophiles. (The use of concentrated frozen suspensions of thermophilic lactic acid bacteria in making Gruyere cheese), *Lait*, 48, 241, 1968.
6. Hynd, J., The use of concentrated single strain cheese starters in Scotland, *J Soc. Dairy Technol.*, 29, 39, 1976.
7. Salji, J. P. and Kroger, M., Effect of using frozen concentrated direct-to the-vat culture on the yield and quality of cheddar cheese, *J. Food Sci.*, 46, 920, 1981.
8. Robinson, R. K., Freeze-dried starter concentrates. I. Their charateristics and potential application to the production of cheese and yogurt, *Dairy Ind. Int.*, 46, 15, 1981.
9. Robinson, R. K., Freeze-dried starter concentrates. II. Their use as direct-to-vat starters in the manufacture of cheddar cheese, *Dairy Ind. Int.*, 46, 16, 1981.
10. Gilliland, S. E., Anna, E. D., and Speck, M. L., Concentrated cultures of *Leuconostoc citrovorum*, *Appl. Microbiol.*, 19, 890, 1970.
11. Gilliland, S. E., Cobb, W. Y., Speck, M. L., and Anna, E. D., Comparison of volatile components produced by concentrated and conventional cultures of *Leuconostoc citrovorum*, *Cult. Dairy Prod. J.*, 6, 12, 1971.
12. Speck, M. L., Interactions among lactobacilli and man, *J. Dairy Sci.*, 59, 338, 1976.
13. Sandine, W. E., Roles of lactobacillus in the intestinal tract, *J. Food Protect.* 42, 259, 1979.

14. Gilliland, S. E., Beneficial interrelationships between certain microorganisms and humans: candiate microorganisms for use as dietary adjuncts, *J. Food Protect.*, 42, 164, 1979.
15. Gilliland, S. E., Preparation and storage of concentrated cultures of lactic streptococci, *J. Dairy Sci.*, 60, 805, 1977.
16. Peebles, M. M., Gilliland, S. E., and Speck, M. L., Preparation of concentrated lactic streptococcus starters, *Appl. Microbiol.*, 17, 805, 1969.
17. Jakubowska, J., Libudzisz, Z., and Piatkiewicz, A., Evaluation of lactic acid streptococci for the preparation of frozen concentrated starter cultures, *Acta Microbiol. Pol.*, 29, 135, 1980.
18. Piatkiewicz, A., Libudzisz, Z., and Jakubowska, J., Evaluation of *Streptococcus cremoris* for preparing frozen concentrated starter cultures, *Acta Microbiol. Pol.*, 26, 407, 1977.
19. Pont, E. G. and Holloway, G. L., A new approach to the production of cheese starter. Some preliminary investigations, *Aust. J. Dairy Technol.*, 23, 22, 1968.
20. Mitchell, S. L. and Gilliland, S. E., Pepsinized sweet whey medium for growing *Lactobacillus acidophilus* for frozen concentrated cultures, *J. Dairy Sci.*, 66, 712, 1983.
21. Lloyd, G. T. and Pont, E. G., An experimental continuous-culture unit for the production of frozen concentrated cheese starters, *J. Dairy Res.*, 40, 149, 1973.
22. Gilliland, S. E., Strain balance of multiple strain lactic streptococcus concentrated cultures, *J. Dairy Sci.*, 54, 1129, 1971.
23. Stadhouders, J., Jansen, L. A., and Hup, G., Preservation of starters and mass production of starter bacteria, *Neth. Milk Dairy J.*, 23, 182, 1969.
24. Stadhouders, J., Hup, G., and Jansen, L. A., A study of the optimum conditions of freezing and storing concentrated mesophilic starters, *Neth. Milk Dairy J.*, 25, 229, 1979.
25. Petterson, H. E., Studies on batch production of bacterial concentrates from mixed species lactic starters, *Appl. Microbiol.*, 29, 133, 1975.
26. Speckman, C. A., Sandine, W. E., and Elliker, P. R., Lyophilized lactic acid starter culture concentrates: preparation and use in inoculation of vat milk for cheddar and cottage cheese, *J. Dairy Sci.*, 57, 165, 1974.
27. Bergire, J. L. and Hermier, J., La production massive de cellules de *Streptocoques lactiquea.* II. Croissance de *Streptococcus lactis* dans un milien A pH constant, *Lait*, 48, 13, 1968.
28. Cogan, T. M., Buckley, D. J., and Condon, S., Optimum growth parameters of lactic streptococci used for the production of concentrated cheese starter culture, *J. Appl. Bacteriol.*, 34, 403, 1971.
29. Efstathiou, J. D., McKay, L. L., Morris, H. A., and Zottola, E. A., Growth and preservation parameters for preparation of a mixed species culture concentrate for cheese manufacture, *J. Milk Food Technol.*, 38, 444, 1975.
30. Bergire, J. L., Production massive de cellules de *Streptocoques lactiques.* III. Production de differentes souches en culture a pH constant, *Lait*, 48, 131, 1968.
31. Lloyd, G. T., New developments in starter technology, *Dairy Sci. Abstr.*, 33, 411, 1971.
32. Keogh, B. P., Survival and activity of frozen starter cultures for cheese manufacture, *Appl. Microbiol.*, 19, 928, 1970.
33. Gilliland, S. E. and Speck, M. L., Biological response of lactic streptococci and lactobacilli to catalase, *Appl. Microbiol.*, 17, 797, 1969.
34. Keen, A. R., Growth studies on the lactic streptococci. III. Observation on continuous growth behavior in reconstituted skim-milk, *J. Dairy Res.*, 39, 151, 1972.
35. Keen, A. R., Growth studies on the lactic streptococci. II. The effect of agitation on the growth characteristics of *Streptococcus lactis* ML 8 in batch culture, *J. Dairy Res.*, 39, 141, 1972.
36. Linklater, P. M. and Griffin, C. J., Growth of *Streptococcus lactis* in milk in a continuous fermentor, *J. Dairy Res.*, 38, 127, 1971.
37. Gilliland, S. E. and Speck, M. L., D-leucine as an auto-inhibitor of lactic streptococci, *J. Dairy Sci.*, 51, 1573, 1968.
38. Lamprech, E. D. and Foster, E. M., The survival of starter organisms in concentrated suspensions, *J. Appl. Bacteriol.*, 26, 359, 1963.
39. Gilliland, S. E. and Speck, M. L., Relationship of cellular components to the stability of concentrated lactic streptococcus cultures at −17°C, *Appl. Microbiol.*, 27, 793, 1974.
40. Lloyd, G. T., The production of concentrated starters by batch culture. II. Studies on the optimum storage conditions, *Aust. J. Dairy Technol.*, 30, 107, 1975.
41. Smittle, R. B., Gilliland, S. E., and Speck, M. L., Death of *Lactobacillus bulgaricus* resulting from liquid nitrogen freezing, *Appl. Microbiol.*, 24, 551, 1972.
42. Goldberg, I. and Eschar, L., Stability of lactic acid bacteria to freezing as related to their fatty acid composition, *Appl. Environ. Microbiol.*, 33, 489, 1977.
43. Smittle, R. B., Gilliland, S. E., Speck, M. L., and Walter, W. M., Jr., Relationship of cellular fatty acid composition to survival of *Lactobacillus bulgaricus* in liquid nitrogen, *Appl. Microbiol.*, 27, 738, 1974.

44. Chapman, H. R., Direct vat inoculation of milk with freeze-dried starters for making cheddar cheese, *J. Soc. Dairy Technol.*, 31, 99, 1978.
45. Nuñez, M., Nuñez, J. A., and Medina, A. L., Freeze-dried concentrated starters for direct inoculation of vat milk in Manchego cheese manufacture, *Milchwissenschaft*, 37, 328, 1982.

Chapter 12

ROLES OF PLASMIDS IN STARTER CULTURES

Larry L. McKay

TABLE OF CONTENTS

I. INTRODUCTION

Plasmid biology is fast becoming an important area of research in bacterial strains used for milk, meat, and plant fermentation processes, as well as for probiotics. Plasmids are defined as autonomous replicating extrachromosomal genetic elements within the bacterial cell. When the cell divides, each daughter cell receives a copy of the chromosomal DNA along with a copy or copies of the plasmids in the parent cell. Since plasmid DNA may replicate independently of the chromosome, any condition resulting in failure of plasmid replication may result in a daughter cell that does not receive a copy of the plasmid. This bacterial cell thus loses the metabolic properties directed by the plasmid DNA. For this reason, plasmid-associated traits may be more unstable than functions controlled by chromosomal genes. Therefore, high spontaneous loss of a metabolic property suggests plasmid DNA involvement. This spontaneous loss is only presumptive evidence, however, and confirmation of the role of plasmids will depend on physical and genetic studies.

In the discussion to follow, the current knowledge with respect to identity and functional properties of plasmids in mesophilic dairy streptococci (*Streptococcus lactis, S. lactis* subsp. *diacetylactis*, and *S. cremoris*), *S. thermophilus*, the lactobacilli, the leuconostocs, and the pediococci will be presented.

II. THE STREPTOCOCCI

A. Mesophilic Streptococci

Metabolic properties such as lactose fermentation, proteinase activity, and citrate utilization are vital for successful dairy fermentations. The instability of these properties in group N streptococci has been known for over 40 years. Little has been done to explain this instability, however, and only in the past 12 years or so, with the advent of plasmid biology and development of techniques for studying genetic composition of dairy streptococci, has it become possible to provide explanations for the instability of these and other metabolic activities. It is now well established that a diversity of plasmid sizes are found within these bacteria.[1-5] Further, it has been demonstrated that these organisms characteristically harbor many plasmid species. The numbers range from 2 to 11, but most strains appear to contain 4 to 7 distinct plasmid sizes. Most of the plasmids observed in these organisms (as well as in other lactic acid bacteria starter cultures) are cryptic (function unknown), but some carry identifiable traits.

At present, more is known about plasmids in the lactic streptococci than in the other members of the lactic acid bacteria used as starter cultures. Some of the metabolic properties observed in lactic streptococci have been clearly established as being plasmid mediated. Examples would be lactose utilization and, in *S. lactis* subsp. *diacetylactis* WM4, the ability to produce a bacteriocin-like substance. Phenotypic and physical evidence for plasmid linkage has been obtained for other traits such as citrate, sucrose, galactose, glucose, mannose and xylose utilization, proteinase activity, modification/restriction systems, and inorganic ion resistance, as well as for nisin production. Further genetic evidence is now needed to confirm plasmid association to these properties. Phenotypic and genetic evidence suggest that diplococcin production by some *S. cremoris* strains may be plasmid-linked. For some characteristics the association with plasmids is highly speculative and is solely based on the phenotypic loss of a metabolic property. In this category would be sensitivity to agglutinins, sensitivity to the lactoperoxidase-thiocyanate-hydrogen peroxide inhibitory system, and exopolymer production. These characteristics will now be briefly reviewed.

1. Lactose Metabolism

Lactose utilization in group N streptococcci is mediated via a phosphoenol-pyruvate (PEP)-dependent phosphotransferase system (PTS).[6] This system appears similar to that described for *Staphylococcus aureus*[7] and is illustrated as follows:

$$\text{PEP} + \text{Enzyme I} \xrightleftharpoons{Mg^{2+}} \text{P-Enzyme I} + \text{Pyruvate}$$

$$\text{P-Enzyme I} + \text{HPr} \rightleftharpoons \text{P-HPr} + \text{Enzyme I}$$

$$\text{P-HPr} + 1/3\ \text{Factor III}^{lac} \rightleftharpoons 1/3\ \text{P-Factor III}^{lac} + \text{HPr}$$

$$1/3\ \text{P-Factor III}^{lac} + \text{Lactose} \xrightarrow[Mg^{2+}]{\text{Enzyme II}^{lac}}$$

$$\text{Lactose-6-P} + 1/3\ \text{Factor III}^{lac}$$

Enzyme I and HPr are soluble cytoplasmic protein components that are synthesized constitutively and are necessary for the transport of many sugars. Factor III^{lac} is a soluble cytoplasmic galactoside-specific component and Enzyme II^{lac} is a membrane-bound component which is galactoside-specific. The latter two components are formed upon induction of the staphylococcal lactose operon, as is phospho-β-galactosidase (P-β-Gal), the enzyme that cleaves lactose-6-P to galactose-6-P and glucose.

The conclusion that lactose-fermenting ability is plasmid mediated in many strains of lactic streptococci is now well established and is based on phenotypic, physical, and genetic evidence. The spontaneous loss of lactose metabolism in these organisms was first noted by early workers[8-13] and later confirmed by McKay et al.[14] The latter workers also found that the incidence of lactose-negative (Lac⁻) variants could be increased by treating the cells with acridine dyes or by culturing at elevated temperatures. The high incidence of spontaneous loss of lactose metabolism coupled with the effect of acridine dyes provided presumptive evidence for the involvement of plasmid DNA in lactose fermentation by lactic streptococci.

Physical evidence for plasmid involvement was provided by comparing the plasmid profiles of lactose-fermenting (Lac⁺) parental cultures with their Lac⁻ variants. Presumptive association of lactose metabolism with plasmid DNA was based on the loss of a plasmid from the Lac⁺ strains. For example, Lac⁻ variants of *S. lactis* C2 were missing a 30 Mdal plasmid;[15,16] a 33, 40, or 45 Mdal plasmid was missing in Lac⁻ *S. lactis* strains ML3, C10, and M18, respectively;[17] Lac⁻ *S. lactis* 11454 (DR1251) lacked a 32 Mdal plasmid;[3] Lac⁻ *S. cremoris* B_1 was missing a 36 Mdal plasmid;[18] and Lac⁻ variants of *S. diacetylactis* 18-16 were missing a 41 Mdal plasmid.[19] Lactose fermentation in *S. lactis* 712 has recently been linked to a 33 Mdal plasmid which is subject to a high rate of deletion formation.[20]

Transduction, conjugation, and transformation have provided the genetic proof that lactose metabolism is plasmid mediated in group N streptococci. Using the temperate phage induced from Lac⁺ *S. lactis* C2 to transduce a plasmid-cured Lac⁻ variant of strain C2, Lac⁺ transductants were obtained possessing a single plasmid of 20 to 23 Mdal.[15] When this plasmid was lost the cells became Lac⁻. Furthermore, the kinetics of UV inactivation of lactose-transducing activity of strain C2 resembled that expected for a plasmid-linked trait.[21] The plasmid in the Lac⁺ transductants was smaller than the 30 Mdal Lac plasmid in strain C2 because the phage head could only accommodate a piece of DNA of about 23 Mdal. Further evidence that this transductionally shortened plasmid coded for lactose utilization came from the observation that protoplasts prepared from Lac⁻ variants of *S. lactis* ML3 were transformed to Lac⁺ by the acquisition of this plasmid.[22] The transfer by transduction of a plasmid associated with lactose metabolism has been reported also for *S. lactis*. 712[1] and for *S. cremoris* C3.[23]

The conjugal transfer of plasmid DNA among group N streptococci has also yielded genetic evidence for linkage of the *lac* genes to plasmid DNA. For example, phenotypic and physical evidence indicated that lactose metabolism in *S. diacetylactis* 18-16 was linked to a 41 Mdal plasmid. Conjugal transfer of the 41 Mdal plasmid from strain 18-16 to a Lac^- plasmid-cured derivative of strain C2 converted the strain to the Lac^+ phenotype.[19] Other studies using conjugal transfer to link *lac* genes with plasmid DNA also have been reported.[24-26] We have found it difficult, at least in contrast with *S. lactis* and *S. diacetylactis*, to obtain phenotypic and physical evidence that lactose metabolism is mediated by plasmid DNA in *S. cremoris*.[27] However, recent speculation suggests that the stability of lactose metabolism in some strains of *S. cremoris* might be due to an unusually stable Lac plasmid since the transfer of the Lac plasmid was observed when several Lac^+ strains of *S. cremoris* were used as donors in conjugal matings.[28]

Although it is now clear that plasmid DNA is involved in the ability of lactic streptococci to utilize lactose, the question arises as to which of the lactose-specific enzymes are coded by plasmid genes. By enzyme complementation analysis it was shown that Lac^- variants of *S. lactis* C2 missing the Lac plasmid were deficient in Enzyme II^{lac}, Factor III,^{lac} and P-β-Gal.[6] Similar results were obtained for Lac^- variants of *S. cremoris* B_1 which lacked the 36 Mdal Lac plasmid. A Lac^- variant of strain B_1 still possessing the Lac plasmid was deficient only in Enzyme II^{lac} activity, indicating a point mutation on the Lac plasmid.[18] These results taken together suggest that all three lactose-specific enzymes are coded by plasmid DNA. Further evidence supporting this conclusion was obtained by St. Martin et al.[29] employing the *Streptococcus sanguis* strain Challis transformation system. To determine which, if any, of the Lac enzymes in *S. lactis* 11454 were coded by the assumed 32 Mdal Lac plasmid, they used purified strain 11454 plasmid DNA to transform an *S. sanguis* Lac^- derivative that lacked the lactose-PTS and P-β-Gal. It was concluded that the lactose plasmid in strain 11454 must code for the synthesis of both the lactose-PTS and P-β-Gal based on the finding that the Lac^+ transformants acquired a lactose-PTS having the same substrate specificity as the *S. lactis* donor strain and because the P-β-Gal in the transformants was the same size as the donor strain enzyme.

With the *lac* genes located on plasmid DNA it has been difficult to isolate single site mutations affecting lactose metabolism. This is probably due to the fact that the frequency of Lac plasmid loss is greater than point mutations on the Lac plasmid, and consequently the Lac^- variants are deficient in the three lactose-specific enzymes. The integration of the Lac plasmid, or a portion of it, into the chromosome of *S. lactis* C2 has now made it possible to isolate Lac^- mutants defective only in Enzyme II^{lac} or Factor II^{lac}.[30,31] Studies on the enzymatic and genetic regulation of lactose metabolism in dairy streptococci should now be greatly facilitated due to the recent advances in our knowledge concerning the Lac plasmid, its integration into the chromosome, and the development of gene transfer systems within this group of organisms.

2. *Galactose Metabolism*

Lac^- variants of *S. lactis* as contrasted with the Lac^+ parental strains, are unable to utilize lactose and exhibit a slower growth rate on galactose, i.e., Lac^- Gal^d.[32-35] This defect in galactose metabolism accompanying the Lac^- phenotype was subsequently explained on the basis that *S. lactis* can utilize galactose by two different metabolic sequences. One involves a PEP-PTS in which galactose enters the cell as galactose-6-phosphate which is converted to tagatose-6-phosphate and subsequently to the triose phosphates, and the other pathway involves an ATP permease system or the Leloir pathway in which galactose enters the cell as free galactose and then is phosphorylated to galactose-1-phosphate.[36] In the *S. lactis* C2 system, as well as in *S. lactis* 11454, it was assumed that galactose was metabolized primarily by the tagatose-6-phosphate pathway in which galactose was translocated into the cell via the lactose-PTS to yield galactose-6 phosphate.[32-34] Lac$^-$ Gal^d derivatives of strains C2 and 11454 were deficient in the lactose-PTS and utilized galactose by the Leloir pathway. These results suggested that the lactose-PTS components carried on the Lac plasmid were also responsible for phosphorylating galactose to galactose-6-phosphate.

Based on physiological studies, Thompson[37] suggests the presence of a PEP-PTS for galactose metabolism separate from the lactose PEP-PTS in *S. lactis*. Genetic evidence for this conclusion was also obtained from our laboratory.[38] Lac$^-$ mutants were isolated from a variant of *S. lactis* C2 in which the lactose plasmid had been integrated into the chromosome. In contrast with the Lac$^-$ Gal^d variants obtained when the lactose plasmid is lost from strain C2, these mutants retained their parental growth characteristics on galactose, i.e., Lac$^-$ Gal$^+$. The Lac$^-$ Gal$^+$ mutants were found to be defective in the lactose-PTS, but still possessed the ability to form galactose-1-phosphate and galactose-6-phosphate. The Lac$^-$ Gal^d mutants formed only galactose-1-phosphate. These results imply that galactose is not translocated via the lactose-PTS, but rather by a specific galactose-PTS for which the genetic locus is also found on the lactose plasmid in *S. lactis* C2. On the other hand, Crow et al.[39] found that the three enzymes of the tagatose-6-phosphate pathway were absent in Lac$^-$ derivatives of *S. lactis* strains C_{10}, H_{11}, and 133 which also had lost their Lac plasmids. Their data suggest that the Lac$^-$ Gal^d phenotype in *S. lactis* is associated with the absence of the three tagatose-6-phosphate enzymes, the lactose-PTS, and P-β-Gal. These enzyme systems were associated with the lactose plasmid in *S. lactis*, but the Lac$^-$ Gal^d cells still retained the ability to form galactose-6-phosphate. Their results imply that galactose is translocated by a galactose-PTS in Lac$^-$ Gal^d cells, the latter phenotype being due to the absence of the tagatose-6-phosphate pathway enzymes.

3. Sugar Fermentations Other than Lactose and Galactose

LeBlanc et al.[3] detected PEP-dependent phosphorylation of lactose, galactose, sucrose, fructose, glucose, and mannose in several strains of *S. lactis*. Two of these strains, 11454 (DR1251) and 354-07 (DR1252), were further examined to determine if the metabolism of any of the above sugars were plasmid mediated. In the presence of acridine orange, sucrose negative (Suc$^-$) variants were isolated at a frequency of 0.03%. These Suc$^-$ isolates were also Lac$^-$, but this was attributed to the high frequency of loss (i.e., 38%) of the Lac$^+$ phenotype under the same curing conditions. Lac$^-$ Suc$^+$ variants were missing a 32 Mdal plasmid, while the Lac$^-$ Suc$^-$ mutants were also missing 28 Mdal plasmid. It therefore appears that sucrose utilization by *S. lactis* 11454 is linked to the 28 Mdal plasmid. It also was found that sucrose-PTS activity existed in Lac$^-$ Suc$^+$ but not in Lac$^-$ Suc$^-$ isolates of strain 11454.

Phenotypic and physical evidence was also obtained linking plasmid DNA to the ability of *S. lactis* 354-07 to utilize glucose and mannose and of *S. lactis* DR1253 to grow on xylose.[3] *S. lactis* 354-07 yielded variants exhibiting poor growth on glucose at a high spontaneous rate (10 to 40%). These variants, which also grew poorly on mannose, were missing a 23 Mdal plasmid. A 3.6 Mdal plasmid was missing in the DR1253 variant that was unable to grow on xylose. These results, coupled with plasmid-mediated lactose and galactose metabolism, suggest that plasmids play a key role in the carbohydrate metabolism of *S. lactis*, *S. cremoris*, and *S. diacetylactis*. Thus, metabolic diversity observed in sugar fermentation patterns among strains within these bacterial groups may be related to plasmid content.

4. Proteolytic Activity

When grown in milk, lactic streptococci are dependent on their proteinase enzyme system to obtain needed nitrogenous compounds from milk proteins. If they lose this proteinase system (Prt⁻), the strains become slow acid producers and require up to 24 hr or longer to coagulate milk at 21°C. A proteinase-positive (Prt⁺) or "fast" culture coagulates milk within 16 to 18 hr at 21°C. Extensive research has been devoted to this proteinase system and it is now known that most of the proteinase activity is associated with the cell wall fraction of the cells.[40,41] It has also been known for many years that Prt⁺ strains may be unstable and spontaneously segregate Prt⁻ variants.[42-45] This conversion from Prt⁺ to Prt⁻ was enhanced when Prt⁺ cells were treated with acridine dyes or grown at elevated temperatures. In 1970 it was first proposed that the loss of plasmid DNA was involved.[46] The first physical evidence for plasmid association with the proteinase system was provided when it was shown by electron microscopy that a Lac⁺ Prt⁻ variant of *S. lactis* C2 lost a 10 Mdal plasmid.[47] Later it was shown by agarose gel electrophoretic techniques that this strain was actually missing a 12.5 and a 18 Mdal plasmid, suggesting that one or both of these plasmids are involved in proteinase production.[16] Prt⁻ variants of *S. lactis* C10[40] and *S. cremoris* HP[48] are missing the cell-wall associated proteinase. In the latter strain this was correlated with the loss of an 8.5 Mdal plasmid.[27] Prt⁻ variants isolated from Prt⁺ *S. cremoris* Wg2 lacked a 16 Mdal plasmid.[49]

While the above phenotypic and physical data implicate the involvement of plasmid DNA in proteinase activity, the question arises as to the nature of the plasmid interrelationship between lactose metabolism and proteinase activity. Strains can become Lac⁺ Prt⁻ or Lac⁻ Prt⁻ and lose different plasmids. For example, Lac⁻ Prt⁻ derivatives of *S. lactis* C2 lack only a 30 Mdal plasmid compared to the parental culture, whereas the Lac⁺ Prt⁻ variant retained the 30 Mdal plasmid but lost the 12.5 and 18 Mdal plasmids.[16,50] This simultaneous loss of Lac⁺ and Prt⁺ phenotypes also has been observed in *S. lactis* strains ML3, C10, and M18 with concomitant loss of a 33, a 40, or a 45 Mdal plasmid, respectively.[17] Similarly, a 31 Mdal plasmid is missing from lactose-defective Prt⁻ variants of *S. diacetylactis* DRC-1.[51] Genetic analysis via transduction supported the conclusion that both Lac and Prt were linked to the 30 Mdal plasmid in strain C2.[15] Why a strain can become Lac⁻ Prt⁻ or Lac⁺ Prt⁻ and lose different plasmids has yet to be resolved and warrants further study. One difficulty is that proteinase activity is not a selectable marker in gene transfer experiments and selection of Prt⁺ recombinants cannot yet be done directly. In addition, the Prt⁺ phenotype may be controlled by several types of proteinases that are coded by different plasmids or by chromosomal DNA. Proteinase activity in *S. cremoris* B_1, for example, appears to be chromosomally mediated.[18] Regulatory control mechanisms operating between plasmid DNA and chromosomal DNA could also affect proteinase activity.

5. Citrate Utilization

The ability of *S. lactis* subsp. *diacetylactis* to produce diacetyl from citrate has led to its widespread use as a flavor producer in many fermented dairy products. Evidence that citrate utilization is plasmid linked has come from several observations. First, it has been reported that *S. diacetylactis* can spontaneously lose the ability to ferment citrate.[52-54] In fact, with the extremely high rate of conversion from citrate positive (Cit⁺) to citrate negative (Cit⁻) observed among some strains coupled with the apparent inability to revert, Kneteman[53] was puzzled as to how Cit⁺ strains could even be found in nature. Second, the frequency of conversion from Cit⁺ to Cit⁻ was increased by the use of plasmid-curing agents.[51] Acridine orange treatment of *S. diacetylactis* strains 18-16 and DRC-1 resulted in the conversion to a Cit⁻ phenotype of 2 and 5%, respectively. The final evidence comes from the correlation of ability to utilize citrate and the presence of 5.5 Mdal plasmids. Absence of the plasmid was associated with the loss of citrate-permease activity. However, these Cit^- variants retained low citritase activity. The correlation between a plasmid of approximately 5.5 Mdal and citrate-fermenting ability also has been noted in other laboratories.[1,56] Since citrate-permease activity appears linked to a 5.5 Mdal plasmid in all strains of *S. diacetylactis* examined, the relationship among the plasmids was determined. Using restriction endonuclease digestions, evidence was provided that the 5.5 Mdal plasmids in the strains examined were identical.[55]

More extensive work is needed in genetic transfer and alteration of genes responsible for citrate utilization. Such research should lead to stabilization of these genes by integration into the chromosome or to strains that have increased ability to utilize citrate and to produce diacetyl.

6. *Antagonistic Properties*

The ability of some group N streptococci to produce inhibitory substances other than organic acids and hydrogen peroxide has been known for some time.[57] These substances include nisin, diplococcin, and bacteriocins. Recent data has implicated the possible involvement of plasmid DNA in the synthesis of these inhibitory agents.

The results of Kozak et al.[58] indicated that stable spontaneous nisin negative (Nis^-) clones occurred at frequencies of 0.03 to 0.82% from some nisin-producing (Nis^+) strains. Growing some Nis^+ strains in the presence of acridine dyes increased the frequency of Nis^- variants to 1.6 and 7.8%. When grown at 40°C the frequency increased to 0.4 and 1.6%. These phenotypic observations certainly implicate the involvement of plasmid DNA. However, attempts to correlate nisin production to plasmid DNA in cesium chloride-ethidium bromide gradients were not conclusive.[59] Some of the Lac^- Nis^- variants were missing the satellite peak characteristic of plasmid DNA, whereas other Lac^- Nis^- mutants possessed the peak. As suggested by Davies and Gasson[1] this ambiguity could reflect the lack of sensitivity in the technique used to determine the presence of plasmid DNA. Davey and Pearce,[60] however, stated they were able to isolate a plasmid-free derivative from Nis^+ *S. lactis* H_1 which retained its ability to produce nisin. They suggested that the *nis* genes of *S. lactis* could be chromosomal or plasmid mediated. In *S. lactis* 11454 (DR1251) there appeared to be an association of the Nis^+ and Suc^+ phenotypes with a 28 Mdal plasmid.[3]

Diplococcin negative (Dip^-) isolates were obtained from several strains of *S. cremoris* at a frequency of 0.1 to 0.2% after growing the Dip^+ parental cultures at 35.5°C.[60] These authors also were unable to detect reversion of the Dip^- isolates to the Dip^+ phenotype, which would suggest the possible involvement of plasmid DNA. The ability to produce diplococcin was transferable by conjugation and Dip^+ transconjugants were able to transfer this ability to *S. lactis* ML_3 and from ML_3 to *S. lactis* C_{10}. However, they were unable to observe any plasmid loss in the Dip^- isolates or the acquisition of plasmid in the Dip^+ transconjugants, which suggested a chromosomal locus for Dip.

Many strains of group N streptococci are capable of producing bacteriocin-like (Bac) substances. Dobrzanski et al.[61] showed that when certain strains were grown under conditions known to eliminate plasmids, Bac^- isolates occurred at about 0.5% frequency. Examination of plasmid profiles were inconclusive as to whether this trait was plasmid mediated. In our laboratory, *S. diacetylactis* WM_4 was shown to exhibit conjugal transfer of the ability to produce a bacteriocin-like substance.[4] Two types of Lac^+ transconjugants were obtained when WM_4 was mated with a plasmid-cured derivative of *S. lactis* C2: Lac^+ Bac^+ and Lac^+ Bac^-. The Lac^+ Bac^+ transconjugants, upon losing the transferred 33 Mdal Lac plasmid, retained the ability to produce Bac (Lac^- Bac^+). In addition, it was subsequently found that Bac-producing ability could be conjugally transferred as a distinct entity, i.e., without co-transfer of the Lac plasmid. Initially our results with Bac closely resembled those observations reported above for Nis and Dip, i.e., we could find no physical evidence that the Bac genes were plasmid mediated. Recently, however, we developed an improved plasmid DNA isolation procedure applicable to lactic streptococci.[62] This procedure allowed us to link Bac-producing ability in strain WM_4 and its transconjugants to a plasmid of approximately 88 Mdal.[63] The question that arises is whether the inability to link plasmid DNA to Nis, Dip, or Bac by the other investigators was also related to the technique used for plasmid isolation. This is possible since the phenotypic data for the trait as well as the conjugal transfer frequencies for Dip certainly implicate plasmid involvement.

7. Restriction-Modification (R/M) Systems

Host-controlled restriction of entering phage DNA has been described as a common bacterial defense mechanism that functions by recognition and degradation of the unmodified DNA.[64] Phage that escape destruction are modified during development to produce lytic phage that are immune to restriction by the host organism. The presence of R/M systems in dairy streptococci was briefly reviewed by Lawrence et al.[65] In 1978, the heterogeneity among strains of lactic streptococci with respect to phage sensitivity was reported and the possibility that R/M systems in lactic streptococci were linked to plasmid DNA was suggested.[66] Loss or inactivation of a restriction system in starved cells of *S. cremoris* KH held at 40°C for 15 min was noted by Pearce.[67] Loss of restriction activity in actively metabolizing cells of *S. cremoris* TR and KH following a heat challenge of 40 to 50°C was also reported.[68] As suggested by Sanders and Klaenhammer,[69] the loss of restriction plasmids during growth at elevated temperatures could generate variants susceptible to phage. To determine whether plasmids play a role in coding for R/M systems, single colony isolates of strain KH were examined for variability of restrictive capacity. Although no completely restrictionless mutants were isolated, from 2 to 6.5% of the colonies isolated were partially deficient in restriction and modification activities. The presence of a 10 Mdal plasmid in strain KH but not in the majority of the examined variants provided physical evidence implicating involvement of plasmids in R/M systems among lactic streptococci. This plasmid linkage provides a genetic mechanism to explain the rapid development of phage-sensitive starter cultures. Since R/M systems provide protection against phage attack, their manipulation by genetic engineering techniques could be one approach for obtaining phage-resistant mutants.[1,69]

8. Drug Resistance

It is somewhat surprising that drug resistance factors have not been readily found in dairy streptococci. Perhaps if isolates from raw milk were examined rather than strains that have been propagated in the laboratory for many years the situation would be different. A recent notable exception is the work by Dobrzanski et al.[61] which indicated that kanamycin resistance (Km^r) is plasmid mediated in certain strains of lactic streptococci. Using transformation, they were able to transfer the Km^r markers from *S. lactis* 71 to *Bacillus subtilis* MT120. From one of the transformants, plasmid DNA was isolated and again transformed into *B. subtilis*, and Km^r clones possessing the plasmid were obtained. This is strong evidence for linkage of drug resistance to plasmid DNA in *S. lactis*. Further studies in this area are now needed to provide selectable plasmid markers applicable to genetic studies in dairy streptococci.

9. Inorganic Ion Resistance

Results obtained from our laboratory indicated that the Lac plasmid from *S. lactis* C2 may harbor the genetic determinants for resistance or sensitivity to inorganic ions.[21]

Lac⁻ derivatives of *S. lactis* C2, missing the 30 Mdal Lac plasmid, were more sensitive to arsenate (Asa), arsenite (Asi), and chromate (Chr), but were more resistant to copper (Cu) than was the Lac⁺ parental culture. Lac⁺ transductants regained their resistance to Asa, Asi, and Chr with the acquisition of the transductionally shortened Lac plasmid; Cu sensitivity was not regained. It was suggested that the genetic locus for the Cu sensitivity trait may be on that segment of plasmid DNA that is lost during the transduction process. Since the resistivity index for the arsenate marker is sufficiently high in *S. lactis* C2, it could possibly be used as a selective trait in recombination experiments. The inorganic ion resistance initially reported for *S. lactic* strains ML3, M18, and C10 needs to be reexamined.[17]

The importance of inorganic ion resistances in dairy fermentations has not been established, but the Cu sensitivity marker may be of significance in certain Swiss-type cheeses in which Cu is believed to play a beneficial role in flavor production.[70] It may be possible to select copper-resistant variants from sensitive strains, allowing a greater variety of usable strains. On the other hand, some strains may carry plasmids determining Cu resistance, as suggested by Otto et al.[49] They found that Prt⁻ variants of *S. cremoris* Wg2 exhibited increased sensitivity to Cu and suggested that a 6.1 Mdal plasmid may determine heavy metal resistance in Prt⁺ *S. cremoris* Wg2.

10. Sensitivity to Agglutinins and to the Lactoperoxidase-Thiocyanate-Hydrogen Peroxide Systems

Raw milk contains a lactoperoxidase-thiocyanate-hydrogen peroxide (LP-SCN-H_2O_2) system in which the SCN in milk is oxidized by LP and H_2O_2 to an intermediate that abruptly halts lactic acid production by sensitive strains of lactic streptococci.[71] Most starters are resistant to this peroxidase system, although they can give rise to sensitive mutants when subcultured in autoclaved milk.[72] The latter workers also showed that when a freeze-dried culture of a resistant strain of *S. lactis* C10 was propagated in glucose broth it yielded 47% resistant cells — 10% of the cells were sensitive to the LP system, and 43% were sensitive to both agglutinins and LP. By subculturing a strain composed of 100% resistant cells under different conditions of medium composition, temperature, and incubation time, it was possible to obtain cultures composed of various proportions of resistant and sensitive cells. Reiter[71] has shown that the strains resistant to the LP system have an enzyme that destroys the inhibitor, whereas the sensitive strains have lost this enzyme. The question that arises is why resistant cultures are unstable with respect to the LP system. It can be speculated that plasmid DNA is responsible for producing the enzyme that allows the resistant strains to overcome the LP inhibitory system. Likewise, the appearance of cells sensitive to agglutinins provides circumstantial evidence that this resistant phenotype may also be plasmid mediated.

11. Exopolymer Production

Slime-forming (ropy) strains of *S. cremoris* were isolated from villi, a fermented milk drink popular in Finland.[73,74] Nonropy variants were obtained by changing the growth temperature from 17 to 30°C. A comparison of the protein profiles between ropy and nonropy variants revealed the presence of certain new bands as well as the elimination and/or decrease in intensity of certain bands in the nonropy variants. The effect of temperature on the appearance of the nonropy variants suggests that plasmid DNA may be associated with this trait. The plasmid could be coding for certain proteins involved in exopolymer production. It is not clear at this time, however, as to the frequency of appearance of the nonropy variants when parental cultures are incubated at 30°C and whether the phenotypic change is stable. It is also worth noting that Hammer[75] in 1923 studied ropiness in cultures of *S. lactis* and questioned the nature of the variability observed in the ropy character of a culture, i.e., nonropy variants could be selected from a presumably pure ropy culture when the latter was plated on an agar medium. Due to the importance of exopolymer production by a variety of strains used in fermented dairy products, the possible relationship between plasmid DNA and the ropy phenotype needs to be examined.

B. *Streptococcus thermophilus*

Unlike the mesophilic dairy streptococci, strains of *S. thermophilus* appear to contain few, if any, plasmid molecules. In a preliminary report Somkuti and Steinberg[76] stated that 26 of 36 strains examined harbored plasmids and that in some strains up to 5 plasmids were detectable. Only 4 of 23 strains examined in our laboratory contained plasmids.[77] Three of these strains possessed a single plasmid species having a molecular weight of less than 2×10^6. The functional properties of plasmids observed in *S. thermophilus* are presently unknown and further research is needed to characterize these plasmids.

III. THE LACTOBACILLI

A. Lactose Metabolism in *Lactobacillus casei*

The first report for plasmids in lactobacilli was by Chassy et al.[78] They reported that *L. casei* subsp. *casei* 64H and *L. casei* subsp. *rhamnosus* each contained a single plasmid having a molecular weight of 23×10^6 and 19×10^6, respectively. Although the function of these plasmids was unknown at the time, they speculated that surface antigens, epithelial adherence sites, or catabolic enzymes could possibly be coded by plasmids present in lactobacilli. Phenotypic evidence for plasmid-mediated lactose metabolism in *L. casei* 1185 was then presented by Hofer.[79] This strain spontaneously lost the ability to ferment lactose and growing the cells in the presence of glucose, at elevated temperatures, or in the presence of acriflavin markedly increased the appearance of Lac^- derivatives. Chassy et al.[80] subsequently provided physical evidence for plasmid-associated lactose metabolism in *L. casei* 64H. They found that two successive subcultures in the presence of acridine dyes and mitomycin C resulted in nearly 100% conversion from Lac^+ to Lac^-, and that the 23 Mdal plasmid found in *L. casei* 64H was consistently absent in the Lac^- clones. These authors also noted that Lac^- strains of 64H exhibited the same phenotypic characteristics as *L. casei* subsp. *alactosus* and thus suggested the possibility that plasmid-borne genes may determine the taxonomic grouping of these subspecies.

L. casei transports lactose and forms lactose-6-phosphate through concomitant phosphorylation via a PEP-PTS. Lactose-6-phosphate is then cleaved by P-β-Gal. The 23 Mdal plasmid in *L. casei* determines both the lactose PEP-PTS and P-β-Gal activities of this strain.[81] Furthermore, Chassy et al.[81] showed that loss of the 15 Mdal plasmid from *L. casei* ATCC 395 and the 36 Mdal plasmid from *L. casei* ATCC 4646 resulted in the loss of lactose metabolism as well as the apparent loss of the structural genes for lactose utilization. This information, coupled with the conjugal transfer of lactose plasmids in *L. casei*,[82] established the plasmid linkage of Lac in this organism. Further proof supporting this conclusion was obtained by Lee et al.[83] through the molecular cloning of the gene determining P-β-Gal of *L. casei* 64H into *Escherichia coli*. Future work involving molecular cloning of the lactose metabolizing genes should lead to a better understanding of the mechanism of enzymatic and genetic regulation of lactose metabolism not only in *L. casei*, but also in other lactobacilli and streptococci important in dairy fermentations.

B. *N*-Acetyl-D-Glucosamine Fermentation in *L. helveticus* subsp. *jugurti*

In 1978, Smiley and Fryder[84] initiated a study with two strains of *L. helveticus* subsp. *jugurti* in an attempt to determine why strain S36-2 produced more lactic acid than strain S13-8. After acriflavin treatment approximately 60% of the colonies isolated from strain S36-2 were unable to ferment *N*-acetyl-D-glucosamine and had reduced lactic acid-producing ability. The level of acid produced was similar to that of strain S13-8. This change of phenotype in strain S36-2 was correlated with the loss of a 20 Mdal plasmid, the sole plasmid present in the *N*-acetyl-D-glucosamine fermenting strain S36-2. Strain S13-8 possessed three cryptic plasmids of approximately 18, 20, and 25 Mdal, respectively. These results provided the first evidence for the presence of plasmids in *L. helveticus* and that one of these plasmids appeared associated with lactic acid production and *N*-acetyl-D-glucosamine fermentation. It would be interesting to determine if this plasmid could stimulate lactic acid production in other lactic acid bacteria.

C. Drug Resistance in *L. fermentum, L. acidophilus,* and *L. reuteri*

The first description of drug resistance plasmids in lactobacilli was by Ishiwa and Iwata.[85] They reported the presence of tetracycline and erythromycin resistance plasmids in a strain of *L. fermentum* isolated from human feces. The plasmids had molecular weights of about 10×10^6 and 37.5×10^6 for tetracycline and erythromycin resistance, respectively. Cryptic plasmids of 8.5 and 16.7 Mdal were observed as well. Vescovo et al.[86] indicated that drug resistance plasmids also existed in *L. acidophilus* and *L. reuteri*. Evidence supporting this conclusion was achieved by curing experiments using acridine dyes and high growth temperatures and by comparing plasmid profiles of antibiotic resistant and sensitive strains. The authors indicated that transformation experiments were being conducted to link drug resistance to specific plasmids.

D. Widespread Occurrence of Plasmids

At present there is considerable interest in the role of *L. acidophilus* in human and animal health. Although all the beneficial effects of this organism have yet to be elucidated, evidence has suggested that some lactobacilli may facilitate resistance to common intestinal disorders through stabilization of the normal intestinal flora.[87-90] Klaenhammer and Sutherland[87] indicated that the intestinal association of naturally occurring strains of *L. acidophilus* suggested that plasmid-mediated antibiotic resistance, bacteriocin production, or colonization factors common to enteric bacteria may also be important to the intestinal activity of the lactobacilli. Only one of eight strains of *L. acidophilus* examined, however, contained plasmid DNA.[87] This strain was an isolate from a pig and possessed two cryptic plasmids of 13.7 and 6.3 Mdal.

Vescovo et al.[91] also found plasmids in *L. acidophilus,* but in only 19% of the examined strains. They also verified the widespread occurrence of plasmids in lactobacilli by examining 195 strains belonging to various species. Plasmids were found in 63% of the strains of *L. reuteri,* 27% of strains of *L. helveticus,* and 4% of strains of *L. casei.* One of 19 strains of *L. bulgaricus* possessed plasmids and no plasmids were detected in 9 strains of *L. plantarum,* 6 strains of *L. coryneformis,* and 5 strains of *L. brevis.* Since various species of lactobacilli are used in dairy, food, and silage fermentation processes, as well as in probiotics, it is essential that research on plasmid biology of these bacteria continue. Plasmids may have a role in their beneficial characteristics.

IV. THE LEUCONOSTOCS

Leuconostoc species are important in a variety of milk and vegetable fermentation processes. This group of bacteria has generally received little attention with respect to plasmid biology. O'Sullivan and Daly,[92] however, examined the plasmid DNA profiles of selected strains of *L. lactis, L. cremoris, L. mesenteroides,* and *L. paramesenteroides.* The number of plasmids per strain varied from 1 to 6 and the sizes ranged from 2.5 to 40 Mdal. Lac^- mutants isolated from *L. mesenteroides* and *L. lactis* were missing a plasmid of 34 and 16 Mdal, respectively. Cit^- variants of *L. mesenteroides* were missing a 10 Mdal plasmid. The characterization of plasmids from *Leuconostoc* species would be a very fruitful area of investigation.

V. THE PEDIOCOCCI

The pediococci, like members from the genus *Leuconostoc,* have yet to be extensively characterized with respect to plasmid content. Species of this group play a role in many food fermentations such as sausage, idli, olive, pickle, sauerkraut, wine, miso, and soy sauce. Plasmid DNA could possibly code for metabolic properties related to the use of these bacteria in food fermentation processes. Graham and McKay[93] have shown that two strains of *Pediococcus cerevisiae* harbored between 4 and 6 plasmids having molecular weights ranging from 4.5×10^6 to 39.5×10^6. Phenotypic and physical data suggested that bacteriocin production was plasmid mediated in one of these strains. Also, two strains of *P. pentosaceus* were examined and found to possess 3 and 5 cryptic plasmids of 4.5 to 34 Mdal.

VI. CONCLUSION

It is now well established that lactic acid bacteria used in starter cultures harbor plasmids of diverse sizes and that some of these plasmids code for properties vital for successful fermentation processes. Attempts to elucidate the functional properties of cryptic plasmids must continue. Amplification of desired plasmid genes should be possible through the isolation of copy number mutants. Such mutants may increase the efficiency of the fermentation process as well as the quality of the final product. Stabilization of plasmid-mediated traits by integrating the genes into the chromosome may also prove beneficial. Clearly, the study of plasmid biology in dairy and food starter cultures has become a prerequisite for future strain improvement programs. This knowledge, coupled with the developing plasmid transfer systems applicable to lactic acid bacteria, is essential for future gene cloning work within this group of organisms.

ACKNOWLEDGMENT

The author wishes to thank K. A. Baldwin for excellent assistance in the preparation of this manuscript.

REFERENCES

1. Davies, F. L. and Gasson, M. J., Reviews of the progress of dairy science: genetics of lactic acid bacteria, *J. Dairy Res.*, 48, 363, 1981.
2. Chopin, A. and Langella, P., Plasmid profiles anologies in group N streptococci, *Lait*, 62, 705, 1982.

3. LeBlanc, D. J., Crow, V. L., and Lee, L. N., Plasmid mediated carbohydrate catabolic enzymes among strains of *Streptococcus lactis*, in *Plasmids and Transposons: Environmental Effects and Maintenance Mechanisms*, Stuttard, C. and Rozee, K. R., Eds., Academic Press, New York, 1980, 31.
4. McKay, L. L. and Baldwin, K. A., Characterization and transferability of plasmids among group N streptococci, in *Microbiology — 1982*, Schlessinger, D., Ed., American Society for Microbiology, Washington, D.C., 1982, 210.
5. Pechmann, H. and Teuber, M., Plasmid pattern of group N (lactic) streptococci, *Zentralbl. Bakteriol. Parasitenkd. Infektionskr. Hyg.*, 1, 33, 1980.
6. McKay, L. L., Regulation of lactose metabolism in dairy streptococci, in *Developments in Food Microbiology*, Vol. 1, Davies, R., Ed., Applied Science Publishers, Essex, England, 1982, 153.
7. Hengstenberg, W., Schrecker, O., Stein, R., and Weil, R., Lactose transport and metabolism in *Staphylococcus aureus*, in *Staphylococci and Staphylococcal Diseases: Proc. 3rd Int. Symp. Staphylococci and Staphylococcal Infect.*, Jeliaszewicz, J., Ed., Gustavo Fischer Verlag, Stuttgart, 1976, 203.
8. Hirsch, A., Growth and nisin production by a strain of *Streptococcus lactis*, *J. Gen. Microbiol.*, 5, 208, 1951.
9. Hunter, G. J. E., Examples of variation within pure cultures of *Streptococcus cremoris*, *J. Dairy Res.*, 10, 464, 1939.
10. Okulitch, O., Microbic dissociation of lactic acid streptococci, *Can. J. Res. Ser. B:*, 17, 17, 1939.
11. Okulitch, O. and Eagles, B. A., Cheese ripening studies. The influence of the configurational relations of the hexoses on the sugar fermenting abilities of lactic acid streptococci, *Can. J. Res. Ser. B:*, 14, 320, 1936.
12. Sherman, J. M. and Hussong, R. V., Fermentative variability among substrains of *Streptococcus cremoris* and *Streptococcus lactis* obtained from pure cultures, *J. Dairy Sci.*, 20, 101, 1937.
13. Yawger, E. S. and Sherman, J. M., Variants of *Streptococcus lactis* which do not ferment lactose, *J. Dairy Sci.*, 20, 83, 1937.
14. McKay, L. L., Baldwin, K. A., and Zottola, E. A., Loss of lactose metabolism in lactic streptococci, *Appl. Microbiol.*, 23, 1090, 1972.
15. McKay, L. L., Baldwin, K. A., and Efstathiou, J. D., Transductional evidence for plasmid linkage of lactose metabolism in *Streptococcus lactis* C2, *Appl. Environ. Microbiol.*, 32, 45, 1976.
16. Klaenhammer, T. R., McKay, L. L., and Baldwin, K. A., Improved lysis of group N streptococci for isolation and rapid characterization of plasmid deoxyribonucleic acid, *Appl. Environ. Microbiol.*, 35, 592, 1978.
17. Kuhl, S. A., Larsen, L. D., and McKay, L. L., Plasmid profiles of lactose-negative and proteinase-deficient mutants of *Streptococcus lactis* C10, ML_3, and M18, *Appl. Environ. Microbiol.*, 37, 1193, 1979.
18. Anderson, D. G. and McKay, L. L., Plasmids, loss of lactose metabolism, and appearance of partial and full lactose-fermenting revertants in *Streptococcus cremoris* B1, *J. Bacteriol.*, 129, 367, 1977.
19. Kempler, G. M. and McKay, L. L., Genetic evidence for plasmid-linked lactose metabolism in *Streptococcus lactis* subsp. *diacetylactis*, *Appl. Environ. Microbiol.*, 37, 1041, 1979.
20. Gasson, M. J., Identification of the lactose plasmid in *Streptococcus lactis* 712, in *Microbiology — 1982*, Schlessinger, D., Ed., American Society for Microbiology, Washington D. C., 1982, 217.
21. Efstathiou, J. D. and McKay, L. L., Inorganic salts resistance associated with a lactose fermenting plasmid in *Streptococcus lactis*, *J. Bacteriol.*, 130, 257, 1977.
22. Kondo, J. K. and McKay, L. L., Transformation of *Streptococcus lactis* protoplasts by plasmid DNA, *Appl. Environ. Microbiol.*, 43, 1213, 1982.
23. Snook, R. J., McKay, L. L., and Ahlstrand, G. G., Transduction of lactose metabolism by *Streptococcus cremoris* C3 temperate phage, *Appl. Environ. Microbiol.*, 42, 897, 1981.
24. McKay, L. L., Baldwin, K. A., and Walsh, P. M., Conjugal transfer of genetic information in group N streptococci, *Appl. Environ. Microbiol.*, 40, 84, 1980.
25. Gasson, M. J. and Davies, F. L., High-frequency conjugation associated with *Streptococcus lactis* donor cell aggregation, *J. Bacteriol.*, 143, 1260, 1980.
26. Walsh, P. M. and McKay, L. L., Recombinant plasmid associated with cell aggregation and high-frequency conjugation in *Streptococcus lactis* ML3, *J. Bacteriol.*, 146, 937, 1981.
27. Larsen, L. D. and McKay, L. L., Isolation and characterization of plasmid DNA in *Streptococcus cremoris*, *Appl. Environ. Microbiol.*, 36, 944, 1978.
28. Snook, R. J. and McKay, L. L., Conjugal transfer of lactose-fermenting ability among *Streptococcus cremoris* and *Streptococcus lactis* strains, *Appl. Environ. Microbiol.*, 42, 904, 1981.
29. St. Martin, E. J., Lee, L. N., and LeBlanc, D. J., Genetic analysis of carbohydrate metabolism in streptococci, in *Microbiology — 1982*, Schlessinger, D., Ed., American Society for Microbiology, Washington, D.C., 1982, 232.

30. McKay, L. L. and Baldwin, K. A., Stabilization of lactose metabolism in *Streptococcus lactis* C2, *Appl. Environ. Microbiol.*, 36, 360, 1979.
31. Wolfe, T. and McKay, L. L., Unpublished data, 1983.
32. Cords, B. R. and McKay, L. L., Characterization of lactose-fermenting revertants from lactose-negative *Streptococcus lactis* C2 mutants, *J. Bacteriol.*, 119, 830, 1974.
33. Demko, G. M., Blanton, S. J. B., and Benoit, R. E., Heterofermentative carbohydrate metabolism of lactose impaired mutants of *Streptococcus lactis*, *J. Bacteriol.*, 112, 1335, 1972.
34. LeBlanc, D. J., Crow, V. L., Lee, L. N., and Garon, C. F., Influence of the lactose plasmid on the metabolism of galactose by *Streptococcus lactis*, *J. Bacteriol.*, 137, 878, 1979.
35. McKay, L. L., Miller, A., III, Sandine, W. E., and Elliker, P. R., Mechanisms of lactose utilization by lactic acid streptococci: enzymatic and genetic analysis, *J. Bacteriol.*, 102, 804, 1970.
36. Bisset, D. L. and Anderson, R. L., Lactose and D-galactose metabolism in group N streptococci: presence of enzymes for both the D-galactose-1-phosphate and D-tagatose-6-phosphate pathway, *J. Bacteriol.*, 117, 318, 1974.
37. Thompson, J., Galactose transport system in *Streptococcus lactis*, *J. Bacteriol.*, 144, 683, 1980.
38. Park, Y. H. and McKay, L. L., Distinct galactose phosphoenolpyruvate-dependent phosphotransferase system in *Streptococcus lactis*, *J. Bacteriol.*, 149, 420, 1982.
39. Crow, V. L., Davey, G. P., Pearce, L. E., and Thomas, T. D., Plasmid linkage of the D-tagatose 6-phosphate pathway in *Streptococcus lactis:* effects on lactose and galactose metabolism, *J. Bacteriol.*, 153, 76, 1983.
40. Thomas, T. D., Jarvis, B. D. W., and Skipper, N. A., Localization of proteinase(s) near the cell surface of *Streptococcus lactis*, *J. Bacteriol.*, 118, 329, 1974.
41. Exterkate, F. A., An introductory study of the proteolytic system of *Streptococcus cremoris* strain HP, *Neth. Milk Dairy J.*, 29, 303, 1975.
42. Citti, J. E., Sandine, W. E., and Elliker, P. R., Comparison of slow and fast acid producing *Streptococcus lactis*, *J. Dairy Sci.*, 48, 14, 1965.
43. Garvie, E. I. and Mabbitt, L. A., Acid production in milk by starter cultures — the effect of peptone and other stimulatory substances, *J. Dairy Res.*, 23, 305, 1956.
44. Harriman, L. A. and Hammer, B. W., Variation in the coagulation and proteolysis of milk by *Streptococcus lactis*, *J. Dairy Sci.*, 14, 40, 1931.
45. Westhoff, D. C., Cowman, R. A., and Speck, M. L., Isolation and partial characterization of a particulate proteinase from a slow acid producing mutant of *Streptococcus lactis*, *J. Dairy Sci.*, 54, 253, 1971.
46. Pearce, L. E., Slow acid variants of lactic streptococci, *Proc. 18th Int. Dairy Congr.*, 1, 118, 1970.
47. McKay, L. L. and Baldwin, K. A., Plasmid distribution and evidence for a proteinase plasmid in *Streptococcus lactis* C2, *Appl. Microbiol.*, 29, 546, 1975.
48. Exterkate, F. A., The proteolytic system of a slow lactic acid producing variant of *Streptococcus cremoris* HP, *Neth. Milk Dairy J.*, 30, 3, 1976.
49. Otto, R., DeVos, W. M., and Gavrieli, J., Plasmid DNA in *Streptococcus cremoris* Wg2: influence of pH on selection in chemostats of a variant lacking a protease plasmid, *Appl. Environ. Microbiol.*, 43, 1272, 1982.
50. McKay, L. L. and Baldwin, K. A., Simultaneous loss of proteinase- and lactose-utilizing enzyme activities in *Streptococcus lactis* and reversal of loss by transduction, *Appl. Microbiol.*, 28, 342, 1974.
51. Kempler, G. M. and McKay, L. L., Characterization of plasmid deoxyribonucleic acid in *Streptococcus lactis* subsp. *diacetylactis:* evidence for plasmid-linked citrate utilization, *Appl. Environ. Microbiol.*, 37, 316, 1979.
52. Collins, E. B. and Harvey, R. J., Failure in the production of citrate permease by *Streptococcus diacetilactis*, *J. Dairy Sci.*, 45, 32, 1962.
53. Kneteman, A., Enrichment and isolation of *Streptococcus citrophilus* van Beynum et Pette, *Antonie van Leeuwenhoek; J. Microbiol. Serol.*, 18, 275, 1952.
54. Mostert, J. F., The ecology and taxonomy of *Streptococcus diacetilactis*, in *Agricultural Research*, Lowes, J., Ed., Department of Technical Services, Pretoria, 1976, 118.
55. Kempler, G. M. and McKay, L. L., Biochemistry and genetics of citrate utilization in *Streptococcus lactis* sp. *diacetylactis*, *J. Dairy Sci.*, 64, 1527, 1981.
56. Kiely, G., Daly, C., and O'Reilly, P., Plasmid DNA in lactic streptcocci, *Ir. J. Food Sci. Technol.*, 6, 99, 1982.
57. Kozak, W., Bardowski, J., and Dobrzanski, W. T., Lactostreptocins — acid bacteriocins produced by lactic streptococci, *J. Dairy Res.*, 43, 247, 1978.
58. Kozak, W., Rajcher-Trzpl, M., and Dobrzanski, W. T., The effect of proflavin, ethidium bromide, and an elevated temperature on the appearance of nisin-negative clones in nisin-producing strains of *Streptococcus lactis*, *J. Gen. Microbiol.*, 83, 295, 1974.

59. Fuchs, P. G., Zajdel, J., and Dobrzanski, W. T., Possible plasmid nature of the determinant for production of the antibiotic nisin in some strains of *Streptococcus lactis, J. Gen. Microbiol.*, 88, 189, 1975.
60. Davey, G. P. and Pearce, L. E., Production of diplococcin by *Streptococcus cremoris* and its transfer to nonproducing group N streptococci, in *Microbiology — 1982*, Schlessinger, D., Ed., American Society for Microbiology, Washington, D.C., 1982, 221.
61. Dobrzanski, W. T., Bardowski, J., Kozak, W., and Zajdel, J., Lactostrepcins: bacteriocins of lactic streptococci, in *Microbiology — 1982*, Schlessinger, D., Ed., American Society for Microbiology, Washington, D.C., 1982, 225.
62. Anderson, D. G. and McKay, L. L., Unpublished data, 1983.
63. Scherwitz, K. M., Baldwin, K. A., and McKay, L. L., Plasmid linkage of a bacteriocin-like substance in *Streptococcus lactis* subsp. *diacetylactis* WM4 and its transferability to *Streptococcus lactis, Appl. Environ. Microbiol.*, 45, 1506, 1983.
64. Arber, W. and Linn, S., DNA modification and restriction, *Annu. Rev. Biochem.*, 38, 467, 1969.
65. Lawrence, R. C., Thomas, T. D., and Terzaghi, B. E., Reviews of the progress of dairy science: cheese starters, *J. Dairy Res.*, 43, 141, 1976.
66. Limsowtin, G. K. Y., Heap, H. A., and Lawrence, R. C., Heterogeneity among strains of lactic streptotocci, *N. Z. J. Dairy Sci. Technol.*, 13, 1, 1978.
67. Pearce, L. E., The effect of host-controlled modification on the replication rate of a lactic streptococcal bacteriophage, *N. Z. J. Dairy Sci. Technol.*, 13, 166, 1978.
68. Sanders, M. E. and Klaenhammer, T. R., Restriction and modification in group N streptococci: effect of heat on development of modified lytic bacteriophage, *Appl. Environ. Microbiol.*, 40, 500, 1980.
69. Sanders. M. E. and Klaenhammer, T. R., Evidence for plasmid linkage of restriction and modification in *Streptococcus cremoris* KH, *Appl. Environ. Microbiol.*, 42, 944, 1981.
70. Kastli, P. O., Recent results of research work in the Swiss federal experimental station for the dairy industry, *J. Soc. Dairy Technol.*, 20, 6, 1967.
71. Reiter, B., Some thoughts on cheese starters, *J. Soc. Dairy Technol.*, 26, 3, 1963.
72. Auclair, J. and Vassal, Y., Occurrence of variants sensitive to agglutinins and to lactoperoxidase in a lactenin-resistant strain of *Streptococcus lactis, J. Dairy Res.*, 30, 345, 1963.
73. Forseen, R. and Pakkila, M., Studies on slime forming group N streptococcus strains. V. Electrophoretic characterization of cell proteins associated with the particulate fraction, *Acta Univ. Oul. A 78 Biochem.*, 22, 1, 1979.
74. Forseen, R., Raunio, V., and Myllymaa, R., Studies on slime forming group N streptococcus strains. I. Differentiation between some lactic streptococcus strains by polyacrylamide gel electrophoresis of soluble cell proteins, *Acta Univ. Oul. A 12 Biochem.*, 3, 1, 1973.
75. Hammer, B. W., Studies on Ropiness in Cultures of *Streptococcus lactis*, Res. Bull. No. 74, Iowa State College of Agriculture, Ames, 1923.
76. Somkuti, G. A. and Steinberg, D. H., Plasmid deoxyribonucleic acid in *Streptococcus thermophilus, J. Dairy Sci.*, 64 (Suppl. 1), 86, 1981.
77. Herman, R. E. and McKay, L. L., Unpublished data, 1983.
78. Chassy, B. M., Gibson, E. M., and Giuffrida, A., Evidence for extrachromosomal elements in *Lactobacillus, J. Bacteriol.*, 127, 1576, 1976.
79. Hofer, F., Involvement of plasmids in lactose metabolism in *Lactobacillus casei* suggested by genetic experiments, *FEMS Microbiol. Lett.*, 1, 167, 1977.
80. Chassy, B. M., Gibson, E. M., and Giuffrida, A., Evidence for plasmid-associated lactose metabolism in *Lactobacillus casei* subsp. *casei, Curr. Microbiol.*, 1, 141, 1978.
81. Chassy, B. M., Gibson, E. M., and Giuffrida, A., Plasmid associated lactose metabolism in *Lactobacillus casei* subsp. *casei, Plasmid*, 2, 296, 1979.
82. Chassy, B. M. and Rokow, E., Conjugal transfer and lactose plasmids in *Lactobacilli casei*, in *Molecular Biology, Pathogenesis, and Ecology of Bacterial Plasmids*, Levy, S., Clowes, R., and Koenig, E., Eds., Plenum Press, New York, 1981, 590.
83. Lee, L., Hansen, J. B., Katarzyna Jagusztyn-Krynicka, E., and Chassy, B. M., Cloning and expression of β-D-phosphogalactoside galactohydrolase gene of *Lactobacillus casei* in *Escherichia coli* K-12, *J. Bacteriol.*, 152, 1138, 1982.
84. Smiley, M. B. and Fryder, V., Plasmids, lactic acid production and N-acetyl-D-glucosamine fermentation in *Lactobacillus helveticus* subsp. *jugurti, Appl. Environ. Microbiol.*, 35, 777, 1978.
85. Ishiwa, H. and Iwata, S., Drug resistance plasmids in *Lactobacillus fermentation, J. Gen. Appl. Microbiol.*, 26, 71, 1980.
86. Vescovo, M., Morelli, L., and Bottazzi, V., Drug resistance plasmids in *Lactobacillus acidophilus* and *Lactobacillus reuteri, Appl. Environ. Microbiol.*, 43, 50, 1982.
87. Klaenhammer, T. R. and Sutherland, S. M., Detection of plasmid deoxyribonucleic acid in an isolate of *Lactobacillus acidophilus, Appl. Environ. Microbiol.*, 39, 671, 1980.

88. Muralidhara, K. S., Sheggeby, G. G., Elliker, P. R., England, D. C., and Sandine, W. E., Effect of feeding lactobacilli on the coliform and lactobacillus flora of intestinal tissue and feces from piglets, *J. Food Protect.*, 40, 288, 1977.
89. Speck, M. L., Interactions among lactobacilli and man, *J. Dairy Sci.*, 59, 338, 1976.
90. Sandine, W. E., Muralidhara, K. S., Elliker, P. R., and England, D. C., Lactic acid bacteria in food and health: a review with special reference to enteropathogenic *Escherichia coli* as well as certain enteric diseases and their treatment with antibiotics and lactobacilli, *J. Milk Food Technol.*, 35, 691, 1972.
91. Vescovo, M., Bottazzi, V., Sarra, P. G., and Dellaglio, F., Evidence of plasmid deoxyribonucleic acid in *Lactobacillus*, *Microbiologica*, 4, 413, 1981.
92. O'Sullivan, T. and Daly, C., Plasmid DNA in *Leuconostoc* species, *Ir. J. Food Sci. Technol.*, 6, 206, 1982.
93. Graham, D. C. and McKay, L. L., Unpublished data, 1983.

Chapter 13

ROLE OF STARTER CULTURE BACTERIA IN FOOD PRESERVATION

Stanley E. Gilliland

TABLE OF CONTENTS

I. HISTORICAL ASPECTS

The original purpose for permitting or encouraging the fermentation of foods was for preservation. The raw product (milk, meats, or vegetables) was allowed to undergo a change to a new form that could be saved for later use. The original processes probably were discovered by accident. The fermentation relied upon microorganisms present in the natural flora of the raw product or in the containers in which the product was stored. The conditions under which the product was stored also would have influenced the type of fermentation. Although at the time the bases for these fermentations was unknown, they provided the origin of many of the cultured foods that we enjoy today.

The advancing development of food processing procedures and cultured foods technology has resulted in the use of starter culture bacteria that initiate the fermentation processes. Starter cultures are selected today more on the basis of the quality of cultured foods they can produce rather than on their ability to preserve the food. Food preservation today is more dependent upon thermal processing and low temperature storage than on fermentation of the food. However, in the recent past, a renewed interest has developed in the preservative actions produced by starter cultures in the effort to find additional means to enhance the preservation of certain food products.

II. PRESERVATIVE ACTION

In most cases, if the starter culture functions properly, the growth of both spoilage and pathogenic microorganisms is retarded during the manufacture and subsequent storage of cultured foods. Perhaps the primary reason for this retardation of growth is the acidic conditions created during the growth of the starter culture bacteria. While the acids produced are perhaps the most important inhibitory substances, evidence has shown that the overall inhibitory action, in many cases, is due to more complex antagonistic systems produced by the starter cultures.[1] These include peroxides, antibiotics, and less well-defined inhibitory metabolites. In all likelihood the overall inhibitory or preservative action produced by starter cultures is due to a combination of factors.

Most of the research studying the mechanisms whereby the preservative actions are produced has utilized cultured milk products. It has been well established that in many of these types of fermentation, the antagonistic action of lactobacilli and streptococci toward both pathogenic and spoilage organisms is due to factors in addition to the acids produced during growth of the culture in the manufacturing process.[2-4] In the manufacture of fermented sausages the acidity that develops due to growth of the lactic acid bacteria inhibits the growth of food-borne pathogens.[5-8] The starter culture bacteria involved in the production of these cultured meat products include lactobacilli, pediococci, and micrococci. Most of the inhibitory action is probably due to action of the lactobacilli and pediococci. However, none of the studies thus far reported have completely described the mechanisms involved in such antagonisms.

The type of acid produced by the starter culture bacteria is very important in regard to the intensity of the antagonistic action they produce. Mather and Babel[9] developed a process for improving the flavor of cottage cheese that involved culturing the cream mixture with *Leuconostoc citrovorum*. In addition to improving the flavor of the product, the shelf-life of the product was extended compared to cheese prepared with an uncultured creaming mixture. The extended shelf-life was due to the inhibition of psychrotrophic spoilage microorganisms that normally grow well under refrigeration conditions. Later studies revealed that the primary inhibitory material produced by the leuconostocs was acetic acid.[10] The inhibition of salmonella by *L. citrovorum* also has been attributed to acetic acid produced during growth of the leuconostocs.[11] Acetic acid is more inhibitory toward these undesirable microorganisms than is an equal concentration of lactic acid. Thus, small amounts of acetic acid produced by starter culture bacteria can make a very important contribution to their preservative actions.

A. Antibiotic Production

Some lactic streptococci and lactobacilli used as starter culture bacteria produce antibiotic compounds.[12,13] Certain strains of *Streptococcus cremoris* produce the antibiotic, diplococcin, that is closely associated with the cells.[14] Its action is variable toward *Staphylococcus aureus* and it exhibits no inhibitory action toward *Escherichia coli*. *S. lactis* produces nisin, an antibiotic that perhaps has received more attention than antibiotics produced by other starter culture bacteria.[15] It is heat stable at acid pH and appears to be a small peptide or protein type compound having a broad spectrum activity, but is primarily active against Gram-positive organisms. *Lactobacillus plantarum* produces lactolin, which differs in characteristics from both diplococcin and nisin.[16] Various strains of *L. acidophilus* produce antibiotics named lactocidin, acidophilin, and acidolin. Lactocidin reportedly is produced only on solid laboratory media and cannot be isolated from liquid media in which the organism had grown.[17] Acidophilin is produced during growth of the lactobacilli in milk and is stable to heat at acid pH.[18] It has a broad spectrum activity against both Gram-positive and Gram-negative microorganisms. Acidolin is somewhat similar to acidophilin in that it is produced by the lactobacilli during growth in milk and it is active against Gram-negative microorganisms.[19] However, apparently, the molecular characteristics of these two antibiotics are different. Lactobacillin, an antibiotic produced by *L. lactis,* has been identified as hydrogen peroxide.[20,21]

Of the antibiotics produced by starter culture bacteria, nisin has received the greatest attention as a food additive to retard the growth of spoilage microorganisms. Because of its heat resistance, it has been suggested that it be added to canned foods prior to the heat process in order to prevent the development of spores that might survive a mild heat treatment.[22] It apparently does not stop spore germination but does prevent outgrowth of the germinated spores. Nisin-producing starter cultures have been used to control "blowing" defects resulting from gas production by clostridia in Swiss-type cheeses.[22]

B. Antagonisms Due to Factors Other than Antibiotics or Acids

Some less well-defined inhibitory metabolites produced during the growth of starter culture bacteria can make significant contributions to the preservative actions of the starter cultures. Some species of bacteria included as starter cultures produce bacteriocins that are antibiotic-like protein compounds.[23,24] Since they are only active against species of bacteria closely related to the producer organisms their role in preservation may be somewhat limited.

Diacetyl produced by citrate fermenting starter culture bacteria (*L. cremoris* and *S. lactis* subsp. *diacetylactis*) may contribute to the ability of these cultures to exert antagonistic action toward undesirable microorganisms. Diacetyl is inhibitory to both food-borne pathogens and spoilage microorganisms.[25,26] It appears to be most effective against the Gram-negative species of bacteria. In the studies in which the inhibitory action of diacetyl has been tested concentrations in excess of 170 ppm were used. Since the level of diacetyl in cultured food products is usually much lower than this, it is not likely to play a major role in the preservative action. However, combined with other metabolites it could be important. Because of its volatility and flavor characteristics it is not likely to be used as an additive to enhance preservation of traditionally noncultured foods.

The inhibition of food-borne pathogens by lactic streptococci is not entirely due to the acid produced by the starter culture bacteria.[27,28] A comparison of the inhibitory

Table 1
INHIBITION OF *SALMONELLA TYPHIMURIUM* AND *STAPHYLOCOCCUS AUREUS* BY LACTIC STREPTOCOCCI IN MILK

Streptococcus culture	Min to reach pH 5.0	Inhibition %[a]	
		S. typhimurium	*S. aureus*
A	351	92.0	98.9
B	360	88.2	98.3
C	375	93.4	98.6
D	381	91.2	98.6
E	387	86.3	98.1
F	435	89.1	98.5

[a] Based on numbers of pathogens in control and associative cultures after 6 hr at 32°C.

From Gilliland, S. E. and Speck, M. L., *J. Milk Food Technol.*, 35, 307, 1972. With permission.

action of six cultures of lactic streptococci toward *Salmonella typhimurium* and *S. aureus* in associative milk cultures is shown in Table 1. The intensity of the antagonistic action did not appear to be directly related to the rapidity of acid production by the streptococci. The six strains of streptococci are listed in the table in order of increasing times required to reach pH 5. The ranking of the percentages of inhibition of the pathogens does not follow the same pattern. Maintaining associative milk cultures of the streptococci and *S. aureus* at pH 6.6 during growth did not eliminate the inhibition. Fractionation and evaluation of whey obtained from milk cultures of inhibitory lactic streptococci revealed that a small molecular weight compound(s) that was neither absorbed by cation nor anion exchange resins was partially responsible for the inhibitory action.[28] Overall, the inhibition produced by the lactic streptococci probably involved a combination of several factors. The antimicrobial activity of *S. thermophilus* is also reportedly due to more than one compound.[29] The inhibitory compounds have been characterized as being of low molecular weight and heat stable with the suggestion that amine compounds are involved.

Some strains of *Pediococcus cerevisiae* isolated from fermenting cucumber brine are inhibitory toward a number of Gram-positive bacteria, but not toward Gram-negative bacteria or yeasts.[30] It has been suggested that this inhibition is not due to acid alone. Furthermore, catalase did not reduce the inhibitory action, thus, hydrogen peroxide apparently was not involved. Pediococci also produce inhibitory action toward bacteria in refrigerated foods. This inhibition has been shown to be due to factors other than acid or hydrogen peroxide.[31] Undoubtedly, the acid produced by these bacteria is partially responsible for the preservative action produced in fermenting vegetables and meat products, however, the overall inhibitory action appears to be due to a combination of factors. Additional studies will be needed to characterize inhibitory metabolites produced by the pediococci.

In vacuum packaged beef stored at 5°C, the numbers of homofermentative lactobacilli have been found to increase while the numbers of the meat spoilage bacteria, *Microbacterium thermosphactum,* did not increase. Based on these observations, it was suggested that an antagonism toward the spoilage organisms was produced by the lactobacilli. Further investigation revealed that the addition of lactobacilli to the meat resulted in reduced growth of *M. thermosphactum* when stored under vacuum packaged conditions, but not when the meat was stored aerobically.[32] It was suggested that the inhibition was due to factors other than lactic acid since the spoilage organism is an acid former and somewhat acid tolerant.

One of the primary metabolites produced by the lactobacilli that contribute to their antagonistic action is hydrogen peroxide. The hydrogen peroxide may itself inhibit undesirable microorganisms or it may react with other components in the foods to form additional inhibitory compounds. For instance, in raw milk hydrogen peroxide may react with thiocyanate and peroxidase to form a series of compounds that are inhibitory to spoilage organisms.[33] It has been suggested that lactobacilli of meat origin produce hydrogen peroxide that reacts with proteinaceous substances in the meat to form inhibitory compounds.[34] Price and Lee[35] observed that during refrigerated storage of oysters, as the number of lactobacilli increased, the number of *Pseudomonas* species decreased. This phenomena was shown to be due to hydrogen peroxide produced by the lactobacilli. Dahiya and Speck[36] reported the antagonistic action of *L. bulgaricus* and *L. lactis* toward *S. aureus* was at least in part due to hydrogen peroxide produced by the lactobacilli. Further, they showed that these lactobacilli accumulated hydrogen peroxide during refrigerated storage without the culture growing and producing acid. Maximum accumulation of hydrogen peroxide occurred at 5°C and at pH 7. The formation of the peroxide by both the lactic streptococci and lactobacilli is greater under aerobic conditions than under anaerobic conditions.[37,38]

III. NEW APPLICATIONS

By understanding the mechanisms whereby the starter culture bacteria inhibit undesirable microorganisms, it may be possible to enhance the preservation action or to apply the inhibitory systems in foods that normally do not contain starter bacteria. In order to achieve the latter, it is of course best if the starter culture bacteria or their inhibitory metabolites do not alter the characteristics of the food product being treated. An example of this is the use of the antibiotic nisin produced by *S. lactis* to inhibit undesirable organisms in cheeses or canned foods.[22]

A. Addition of Spent Media to Foods

Several researchers have suggested the addition of spent media from starter cultures to foods as preservatives.[39-42] Variable results from such attempts have been reported. Cell-free whey obtained from milk cultures of *S. thermophilus* did not inhibit *E. coli*, *P. fragi*, *Micrococcus flavus*, or *S. aureus* when added to milk cultures. However, whey from a milk culture of *L. acidophilus* inhibited these undesirable organisms.[39] In a similar study, material extracted by methanol from a milk culture of *S. thermophilus* and added to refrigerated ground beef was inhibitory to spoilage organisms.[41] However, the inhibitory action noted in this study may have been related to acid compounds in the extract since the pH of the meat decreased as the concentration of the extract was increased. This inhibitory system was more active in meat than in milk.

The success of such an approach to using inhibitory metabolites must be dependent on the capability of the individual strains of the various species of starter bacteria to produce the inhibitory compounds. With the variation known to exist among strains of starter culture bacteria with regard to the activity related to the production of cultured foods, variation must certainly exist with respect to the intensity of inhibitory actions produced. Most of the studies in which spent media or fractions of spent media have been tested included only one strain of the species of starter culture being evaluated. It seems reasonable that with a system of selecting starter culture bacteria for the ability to produce inhibitory substances that it may be possible to select those strains that are more potent and effective in inhibiting undesirable microorganisms. Additionally, the composition of the medium in which the starter culture bacteria are grown may influence the intensity of the inhibitory system.[42]

B. Addition of Cells of Starter Culture Bacteria to Refrigerated Foods

Our advancing technology which has led to improved refrigeration systems for storing foods has in some cases created new problems with regard to food spoilage. The low temperature maintained by these refrigeration systems creates an environment that favors the growth of psychrotrophic spoilage microorganisms. Growth of these microorganisms not only reduces the quality of the foods due to their growth and action, but some produce proteolytic and/or lipolytic enzymes that are extremely heat resistant. If formed in the raw food during refrigerated storage these enzymes can survive the heat process and cause spoilage defects in the heat processed food. Thus, control of psychrotrophic microorganisms in raw foods becomes even more important. Since these organisms grow well in refrigeration it is desirable to have some supplementary control system in order to prevent their growth. The fact that some starter culture bacteria, e.g., the lactobacilli, can produce inhibitory compounds such as hydrogen peroxide at refrigeration temperature without growing has led to suggestions for their use in aiding the control of psychrotrophic microorganisms.[35,36]

Addition of lactic cultures (lactic streptococci with or without *Leuconostoc* species) to various refrigerated foods has been shown to reduce the growth of spoilage microorganisms.[43-45] The addition of lactic culture, which had been grown in milk, to ground beef at a level of 10% inhibited the growth of Gram-negative bacteria during storage of meat at 7°C.[43] However, with the amount added some influence on the flavor of the meat must have resulted. The mechanisms whereby this antagonistic action was produced was not reported. The inhibitory action of cultures of *S. diacetylactis* when added to milk and meat products has been suggested to be due to volatile fatty acids produced to the organism.[44]

Juffs and Babel[45] added lactic culture grown in milk at a level of 0.5% to raw milk which was subsequently stored at 3.7 or 7°C in an effort to control the growth of psychrotrophic bacteria in the milk. The lactic culture retarded the growth of psychrotrophic spoilage microorganisms at both temperatures (Table 2). Greater inhibition was observed at 3.5 than at 7°C. Hydrogen peroxide was identified as being, at least partially, responsible for the inhibition. They further reported variation among the lactic starters with regard to the ability to produce inhibition and the intensity of the inhibitory action. There was only a slight decrease in pH of the milk during the refrigerated storage period. The lactic cultures were more active in heated milk that had been inoculated with pure cultures of psychrotrophs than against the natural flora in raw milk.

Cells of *P. cerevisiae* added to meat and to pasteurized liquid whole eggs retarded the growth of psychrotrophic spoilage organisms during refrigerated storage.[31,46,47] The mechanisms whereby the pediococci exert such inhibitory action has not been elucidated; however, it has been established that it is not due to hydrogen peroxide.[31] The determination of the mechanisms whereby this action occurs may lead to a more effective use of cells of this starter culture to aid in the control of the growth of psychrotrophic microorganisms in refrigerated foods.

Because species of lactobacilli produce more hydrogen peroxide than do other species of starter culture bacteria, they perhaps offer the best possibility for exerting antagonistic actions at refrigeration temperatures.[31] The extent of the antagonistic action produced by lactobacilli varies among species and strains within species. Antagonisms toward psychrotrophic bacteria have been reported to occur when cells of lactobacilli are added to refrigerated milk, meat, pasteurized eggs, and seafood.[31,46-49] Results

Table 2
GROWTH OF PSYCHROTROPHIC BACTERIA IN RAW MILKS AT 7°C OR AT 7 AND 3.5°C IN THE PRESENCE AND ABSENCE OF LACTIC CULTURE NO. 44

Sample no.	Storage temp. (°C)	Lactic culture added	Psychrotrophic count/mℓ[a] after storage for: 0 hr	24 hr	48 hr	72 hr
1	7	−	43×10^2	nd[b]	41×10^4	66×10^5
		+	100%	nd	35%	28%
2	7	−	110×10^3	168×10^4	218×10^5	96×10^6
		+	100%	73%	78%	96%
3	7	−	88×10^4	128×10^5	96×10^6	nd
		+	100%	100%	83%	nd
	3.5	−	88×10^4	33×10^5	250×10^5	nd
		+	100%	91%	68%	nd
4	7	−	12×10^2	nd	80×10^3	nd
		+	100%	nd	50%	nd
	3.5	−	12×10^2	nd	62×10^2	nd
		+	100%	nd	52%	nd

[a] Total psychrotrophic counts are given for the control samples whereas counts of psychrotrophic bacteria in samples with lactic culture added are given as the percentage of the control count for the corresponding storage period.
[b] nd = Not determined.

From Juffs, H. S. and Babel, F. J., *J. Dairy Sci.*, 58, 1612, 1975. With permission.

Table 3
EFFECT OF CELLS OF *L. BULGARICUS, L. LACTIS,* AND *P. CEREVISIAE* ON GROWTH OF PSYCHROTROPHIC BACTERIA IN GROUND BEEF[a] AT 5°C

Days at 5°C	Control	*L. bulgaricus*[b] NCS1	*L. lactis*[b] BYL1	*P. cerevisiae*[b]
0	6.7×10^2 [c]	8.3×10^2	8.5×10^2	8.4×10^2
3	1.5×10^5	5.3×10^3	4.7×10^3	1.5×10^4

[a] Ground beef prepared in laboratory.
[b] Approximately 5×10^8 cells added per gram.
[c] Colony counts per gram on CVT agar.

Reprinted from *J. Food Sci.*, 1975, 40, 903. Copyright © by Institute of Food Technologists. With permission.

from a recent study revealed that additional cells of *L. plantarum* to scallops stored at 4°C had no effect on spoilage of the product.[50] This is in contrast to a number of other studies that have shown that other species of lactobacilli added to refrigerated foods do exert inhibitory actions. The difference may be due to the use of too few cells of the *L. plantarum* or to the fact that *L. plantarum* was not the best choice for such use. Cells of *L. bulgaricus* added to ground beef exert less inhibitory action than do cells of *L. lactis.*[31] Both of these lactobacilli, however, were more inhibitory than cells of *P. cerevisiae* (Table 3). In related experiments, the addition of 8×10^7 cells of *L. bulgaricus* per milliliter to refrigerated autoclaved milk that had been inoculated with *P. fragi*

Table 4
INHIBITION OF GROWTH OF PSYCHROTROPH RM IN AUTOCLAVED 10% NFMS BY CELLS OF DIFFERENT CULTURES OF *LACTOBACILLUS BULGARICUS*[a]

	Inhibition (%)			
Culture	Trial 1	Trial 2	Trial 3	Av.
L. bulgaricus D1	48	99	99	82
L. bulgaricus D3	62	91	75	76
L. bulgaricus E2	30	83	40	51
L. bulgaricus F1	−48	77	49	26
L. bulgaricus F2	48	83	32	54

[a] 2.5×10^7 lactobacilli added/m*l* for each culture; initial population of psychrotroph: 1×10^3/m*l*.

From Martin, D. R. and Gilliland, S. E., *J. Food Protect.*, 43, 675, 1980. With permission.

resulted in a bactericidal action toward the *Pseudomonas* species. The inhibitory action exerted by *L. bulgaricus* and *L. lactis* was due primarily to hydrogen peroxide since the addition of active catalase inhibited the inhibitory action.[31] Little or no change was observed in the pH of either the meat or milk during refrigerated storage at 5°C. Variations in the amount of hydrogen peroxide produced by different species of lactobacilli suggest that some may be more inhibitory than others.[31,51]

Cells of *L. bulgaricus*, when added to autoclaved milk that had been inoculated with a psychrotrophic spoilage organism isolated from raw milk resulted in inhibition of the psychrotroph during storage of the milk at 5.5°C.[48] The *L. bulgaricus* was added to the milk at a level of 2.5×10^7 cells per milliliter. Six strains of *L. bulgaricus* isolated from commercial yogurt were compared for their ability to inhibit the psychrotrophic culture (Table 4). After 6 days of storage at 5.5°C the average inhibition of the psychrotrophic culture ranged from 26 to 82%. No decrease was noted in the pH of any of the milk samples. In general, the strains of *L. bulgaricus* that produced the highest levels of hydrogen peroxide also produced the greatest levels of inhibition. Related experiments revealed that as the numbers of lactobacilli were increased the intensity of the inhibition also increased. While all six strains of *L. bulgaricus* exerted an inhibitory action in the heated milk, none were able to inhibit the growth of psychrotrophic microorganisms in raw milk during refrigerated storage. This was true even when the lactobacilli were added to the milk at a level of 1×10^8 cells per milliliter.

Inhibition of the growth of psychrotrophic microorganisms in raw milk during refrigerated storage has been accomplished by adding cells of *L. lactis*.[49] Six strains of *L. lactis* were compared for their inhibitory action (Table 5). In general, as the amount of hydrogen peroxide production increased, so did the intensity of the antagonistic action. Previous reports had indicated that *L. lactis* generally produces more hydrogen peroxide than does *L. bulgaricus*.[31,51] This probably explains why cells of *L. lactis* were effective in raw milk while cells of *L. bulgaricus* were not. Presumably, more hydrogen peroxide would be required to exert an antagonistic action in raw milk than in heated milk since the raw milk would contain active catalase and peroxidase that could degrade the peroxide as it is produced. In the study comparing the strains of *L. lactis*, it was also reported that the addition of 0.1 and 0.2% potassium sorbate to the raw milk

Table 5
NUMBERS OF NONLACTOBACILLI IN RAW MILK WITH AND WITHOUT ADDED CELLS OF *LACTOBACILLUS LACTIS*[a] AFTER 4 DAYS OF STORAGE AT 5°C

Sample	Days at 5°C	Nonlactobacilli/mℓ[b] Trial 1	Trial 2	Trial 3	Trial 4	Trial 5	Trial 6	Av.
Control	0	3.72	3.18	3.34	3.45	3.82	3.60	
Control	4	4.20	3.61	7.23	5.40	4.58	7.38	5.40[c]
L. lactis B	4	4.00	3.34	5.85	5.40	4.20	6.76	4.93[d]
L. lactis Farr	4	4.04	3.56	5.79	5.72	4.59	6.69	5.07[c]
L. lactis 12315	4	3.96	3.77	5.92	5.23	4.53	6.70	5.02[c]
L. lactis 403E-15	4	4.08	3.20	5.69	5.76	4.67	6.68	5.01[c]
L. lactis 39A1	4	4.15	3.15	5.77	3.89	4.41	6.57	4.66[d]
L. lactis 39A2	4	4.18	3.57	5.79	5.79	4.54	6.91	5.13[c]

[a] *L. lactis* counts were approximately 1×10^8/mℓ.
[b] Counts recorded as $\log_{10}$ of numbers of nonlactobacilli/mℓ.
[c,d] Counts followed by different letters significantly differ (strain 39A1 at $p < 0.01$ and strain B at $p < 0.05$).

From Gilliland, S. E. and Ewell, H. R., *J. Dairy Sci.*, 66, 974, 1983. With permission.

in combination with cells of the lactobacilli exerted greater antagonistic action than either the lactobacilli or sorbate alone.[49] The results indicated that sorbate did not enhance peroxide production by the lactobacilli. The mechanisms whereby this apparent synergistic action occurs is not known, however, it may be merely an additive effect. The numbers of cells of *L. lactis* used in these experiments were high (1×10^8/mℓ), which would make such an endeavor a relatively expensive treatment for helping control the growth of psychrotrophs in raw milk. However, through more thorough strain selection, cultures may be discovered that are more active in producing inhibitory systems and thus fewer cells would be required for controlling the psychrotrophs. Once we better understand the factors involved in the production of hydrogen peroxide by these cells at refrigeration temperatures we may be able to produce cells of the lactobacilli under conditions that would enhance their ability to exert inhibitory action in refrigerated foods.

IV. FUTURE DIRECTIONS

Additional research is needed in order to more clearly define and characterize inhibitory systems produced by the starter culture bacteria. Rapid methods for measuring the production of inhibitory metabolites or actions would improve the screening of cultures for possible use in inhibiting growth of undesirable microorganisms. It may be that combinations of the inhibitory systems produced by starter culture bacteria with other control measures may prove to be more efficient preservation systems. It seems unlikely that the use of starter culture bacteria will replace the highly efficient refrigeration systems and heat processing treatments for preserving foods. However, the use of cells of starter culture bacteria and/or metabolites produced by them may be very useful as adjuncts in enhancing the control of psychrotrophic microorganisms in refrigerated foods.

REFERENCES

1. Hurst, A., Microbial antagonisms in foods, *Can. Inst. Food Sci. Technol. J.*, 6, 80, 1973.
2. Park, H. S. and Marth, E. H., Behavior of *Salmonella typhimurium* in skim milk during fermentation by lactic acid bacteria, *J. Milk Food Technol.*, 35, 482, 1972.
3. Reiter, B., Fewins, B. G., Fryer, T. F., and Sharpe, M. E., Factors affecting the multiplication and survival of coagulase positive staphylococci in Cheddar cheese, *J. Dairy Res.*, 31, 261, 1964.
4. Ross, G. D., The inhibition of growth of spoilage microorganisms in milk by *Streptococcus lactis* subsp. *diacetylactis, Leuconostoc cremoris* and *L. dextranicum*, *Aust. J. Dairy Technol.*, 36, 147, 1981.
5. Goepfert, J. M. and Chung, K. C., Behavior of *Salmonella* during the manufacture and storage of a fermented sausage product, *J. Milk Food Technol.*, 33, 185, 1970.
6. Smith, J. L., Huhtanen, C. N., Kissinger, J. C., and Palumbo, S. A., Survival of salmonellae during pepperoni manufacture, *Appl. Microbiol.*, 30, 759, 1975.
7. Niskanen, A. and Nurmi, E., Effect of starter culture on staphylococcal enterotoxin and thermonuclease production in dry sausage, *Appl. Environ. Microbiol.*, 31, 11, 1976.
8. Daly, C., La Chance, M., Sandine, W. E., and Elliker, P. R., Control of *Staphylococcus aureus* in sausage by starter cultures and chemical acidulation, *J. Food Sci.*, 38, 426, 1973.
9. Mather, D. W. and Babel, F. J., Inhibition of certain types of bacterial spoilage in creamed cottage cheese by the use of a creaming mixture prepared with *Streptococcus citrovorus*, *J. Dairy Sci.*, 42, 1917, 1959.
10. Pinheiro, A. J. R., Liska, B. J., and Parmelee, C. E., Properties of substances inhibitory to *Pseudomonas fragi* produced by *Streptococcus citrovorus* and *S. diactilactis*, *J. Dairy Sci.*, 51, 183, 1968.
11. Sorrells, K. M. and Speck, M. L., Inhibition of *Salmonella gallinarum* by culture filtrates of *Leuconostoc citrovorum*, *J. Dairy Sci.*, 53, 239, 1970.
12. Babel, F. J., Antibiosis by lactic culture bacteria, *J. Dairy Sci.*, 60, 815, 1976.
13. Speck, M. L., Interactions among lactobacilli and man, *J. Dairy Sci.*, 59, 338, 1976.
14. Oxford, A. E., Diplococcin, an antibacterial protein elaborated by certain milk streptococci, *Biochem. J.*, 38, 178, 1944.
15. Hurst, A., Biosynthesis of the antibiotic nisin and other basic peptides by *Streptococcus lactis* grown in batch culture, *J. Gen. Microbiol.*, 45, 503, 1966.
16. Kodama, R., Studies on lactic acid bacteria. II. Lactolin, a new antibiotic substance produced by lactic acid bacteria, *J. Antibiot.*, 5, 72, 1952.
17. Vincent, J. G., Veomett, R. C., and Riley, R. F., Antibacterial activity associated with *Lactobacillus acidophilus*, *J. Bacteriol.*, 78, 477, 1959.
18. Vakil, J. R. and Shahani, K. M., Partial purification of antibacterial activity of *Lactobacillus acidophilus*, *Bacteriol. Proc.*, p. 9, 1965.
19. Hamdan, I. Y. and Mikolajcik, E. M., Acidolin: an antibiotic produced by *Lactobacillus acidophilus*, *J. Antibiot.*, 27, 631, 1974.
20. Wheater, D. M., Hirsch, A., and Mattick, A. T. R., "Lactobacillin" an antibiotic from lactobacilli, *Nature (London)*, 168, 659, 1951.
21. Wheater, D. M., Hirsch, A., and Mattick, A. T. R., Possible identity of "Lactobacillin" with hydrogen peroxide produced by lactobacilli, *Nature (London)*, 170, 623, 1952.
22. Hurst, A., Interactions of food starter cultures and food-borne pathogens: the antagonisms between *Streptococcus lactis* and sporeforming microbes, *J. Milk Food Technol.*, 35, 418, 1972.
23. Upreti, G. C. and Hinsdill, R. D., Production and mode of action of lactocin 27: bacteriocin from a homofermentative *Lactobacillus*, *Antimicrob, Agents Chemother.*, 7, 139, 1975.
24. Barefoot, S. F. and Klaenhammer, T. R., Detection and activity of lactacin B, a bacteriocin produced by *Lactobacillus acidophilus*, *Appl. Environ. Microbiol.*, 45, 1808, 1983.
25. Gupta, K. G., Chandiok, L., and Bhatnagar, L., Antibacterial activity of diacetyl and its influence on the keeping quality of milk, *Zentralbl. Bakteriol. Parasitenkd. Infektionskr. Hyg. Abt. 1: Orig. Reihe B:*, 158, 202, 1973.
26. Jay, J. M., Antimicrobial properties of diacetyl, *Appl. Environ. Microbiol.*, 44, 525, 1982.
27. Gilliland, S. E. and Speck, M. L., Interactions of food starter cultures and food-borne pathogens: lactic streptococci versus staphylococci and salmonellae, *J. Milk Food Technol.*, 35, 307, 1972.
28. Gilliland, S. E. and Speck, M. L., Antagonism of lactic streptococci toward *Staphylococcus aureus* in associative milk cultures, *Appl. Microbiol.*, 28, 1090, 1974.
29. Pulsusani, S. R., Rao, D. R., and Sunki, G. R., Antimicrobial activity of lactic cultures: partial purification and characterization of antimicrobial compound(s) produced by *Streptococcus thermophilus*, *J. Food Sci.*, 44, 575, 1979.
30. Fleming, H. P., Etchells, J. L., and Costilow, R. N., Microbial inhibition by an isolate of *Pediococcus* from cucumber brine, *Appl. Microbiol.*, 30, 1040, 1975.

31. Gilliland, S. E. and Speck, M. L., Inhibition of psychrotrophic bacteria by lactobacilli and pediococci in nonfermented refrigerated foods, *J. Food Sci.*, 40, 903, 1975.
32. Roth, L. A. and Clark, D. S., Effect of lactobacilli and carbon dioxide on the growth of *Microbacterium thermosphactum* on fresh beef, *Can. J. Microbiol.*, 21, 629, 1975.
33. Björck, L., Claesson, O., and Schulthess, W., The lactoperoxidase/thiocyanate/hydrogen peroxide system as a temporary preservative for raw milk in developing countries, *Milchwissenschaft*, 34, 726, 1979.
34. Talon, R., Labadie, J., and Larpent, J. P., Characterization of the inhibitory power of *Lactobacillus* of meat origin, *Zentralbl. Bakteriol. Parasitenkd. Infektionskr. Hyg. Abt. 1: Orig. Reihe B:*, 170, 133, 1980.
35. Price, R. J. and Lee. J. S., Inhibition of *Pseudomonas* species by hydrogen peroxide producing lactobacilli, *J. Milk Food Technol.*, 33, 13, 1970.
36. Dahiya, R. S. and Speck, M. L., Hydrogen peroxide formation by lactobacilli and its effects on *Staphylococcus aureus*, *J. Dairy Sci.*, 51, 1568, 1968.
37. Gilliland, S. E. and Speck, M. L., Biological response of lactic streptococci and lactobacilli to catalase, *Appl. Microbiol.*, 17, 797, 1969.
38. Collins, E. B. and Aramaki, K., Production of hydrogen peroxide by *Lactobacillus acidophilus*, *J. Dairy Sci.*, 63, 353, 1980.
39. Singh, J., Inhibition of growth of spoilage microorganisms by *Streptococcus thermophilus* and *Lactobacillus acidophilus* in cow, buffalo and goat milk, *J. Food Protect.*, 46, 497, 1983.
40. Branen, A. L., Go, H. C., and Genske, R. P., Purification and properties of antimicrobial substances produced by *Streptococcus diacetilactis* and *Leuconostoc citrovorum*, *J. Food Sci.*, 40, 446, 1975.
41. Rao, D. R., Reddy, B. M., Sunki, G. R., and Pulsusani, S. R., Influence of antimicrobial compound(s) extracted from milk fermented by *Streptococcus thermophilus* on keeping quality of meat and milk, *J. Food Qual.*, 4, 247, 1982.
42. Reddy, N. S and Ranganathan, B., Nutritional factors affecting growth and production of antimicrobial substances by *Streptococcus lactis* subsp. *diacetylactis* S_1-67/C, *J. Food Protect.*, 46, 514, 1983.
43. Reddy, S. G., Henrickson, R. L., and Olson, H. C., The influence of lactic cultures on ground beef quality, *J. Food Sci.*, 35, 787, 1970.
44. Daly, C., Sandine, W. E., and Elliker, P. R., Interactions of food starter cultures and food-borne pathogens: *Streptococcus diacetilactis* versus food pathogens, *J. Milk Food Technol.*, 35, 349, 1972.
45. Juffs, H. S. and Babel, F. J., Inhibition of psychrotrophic bacteria by lactic cultures in milk stored at low temperature, *J. Dairy Sci.*, 58, 1612, 1975.
46. Raccach, M., Baker, R. O., Regenstein, J. M., and Mulnix, E. J., Potential application of microbial antagonism to extended storage stability of a flesh type food, *J. Food Sci.*, 44, 43, 1979.
47. Raccach, M. and Baker, R. D., Bacterial repressive action of meat starter culture in pasteurized liquid whole egg, *J. Food Sci.*, 44, 90, 1979.
48. Martin, D. R. and Gilliland, S. E., Inhibition of psychrotrophic bacteria in refrigerated milk by lactobacilli isolated from yogurt, *J. Food Protect.*, 43, 675, 1980.
49. Gilliland, S. E. and Ewell, H. R., Influence of combinations of *Lactobacillus lactis* and potassium sorbate on growth of psychrotrophs in raw milk, *J. Dairy Sci.*, 66, 974, 1983.
50. Bremner, H. A. and Statham, J. A., Spoilage of vacuum-packed chill-stored scallops with added lactobacilli, *Food Technol. Aust.*, 35, 284, 1983.
51. Premi, L. and Bottazzi, V., Hydrogen peroxide formation and hydrogen peroxide splitting activity in lactic acid bacteria, *Milchwissenschaft*, 27, 762, 1972.

INDEX

A

B

C

D

E

F

G

H

I

J

K

L

M

Q

R

S

T

U

V

W

X

Y

Z